The (Burning) Case for a GREEN NEW DEAL

NAOMI KLEIN

SIMON & SCHUSTER

NEW YORK LONDON TORONTO SYDNEY NEW DELHI

Simon & Schuster
1230 Avenue of the Americas
New York, NY 10020

First Simon & Schuster hardcover edition September 2019

Interior design by Ruth Lee-Mui

Manufactured in the United States of America.

1 3 5 7 9 10 8 6 4 2

Library of Congress Cataloging-in-Publication Data is available.

ISBN 978-1-9821-2991-0
ISBN 978-1-9821-2993-4 (ebook)

For

ARTHUR MANUEL
1951–2017

The future isn't cast into one inevitable course. On the contrary, we could cause the sixth great mass extinction event in Earth's history, or we could create a prosperous civilization, sustainable over the long haul. Either is possible starting from now.

—KIM STANLEY ROBINSON

CONTENTS

INTRODUCTION: "WE ARE THE WILDFIRE"

ON A FRIDAY IN MID-MARCH 2019, THEY STREAMED OUT OF SCHOOLS IN LITTLE rivulets, burbling with excitement and defiance at an illicit act of truancy. The little streams emptied onto grand avenues and boulevards, where they combined with other flows of chanting and chatting children and teens, dressed in leopard leggings and crisp uniforms and everything in between.

Soon the rivulets were rushing rivers: 100,000 bodies in Milan, 40,000 in Paris, 150,000 in Montreal.

Cardboard signs bobbed above the surf of humanity: THERE IS NO PLANET B! DON'T BURN OUR FUTURE. THE HOUSE IS ON FIRE!

Some placards were more intricate. In New York City, a girl held up a lush painting of delicate bumble bees, flowers, and jungle animals. From a distance, it looked like a school project on biodiversity; up close, it was a lament for the sixth mass extinction: 45% OF INSECTS LOST TO CLIMATE CHANGE. 60% OF ANIMALS HAVE

DISAPPEARED IN THE LAST 50 YEARS. At the center she had painted an hourglass rapidly running out of sand.

For the young people who participated in the first ever global School Strike for Climate, learning has become a radicalizing act. In early readers, textbooks, and big-budget documentary films, they learned of the existence of ancient glaciers, dazzling coral reefs, and exotic mammals that make up our planet's many marvels. And then, almost simultaneously—from teachers, older siblings, or sequels to those same films—they discovered that much of this wonder has already disappeared, and much of the rest of it will be on the extinction block before they hit their thirties.

But it wasn't only learning about climate change that moved these young people to march out of class en masse. For a great many of them, it was also living it. Outside the legislature building in Cape Town, South Africa, hundreds of young strikers chanted at their elected leaders to stop approving new fossil fuel projects. It was just one year ago that this city of four million people was in the clutches of such severe drought that three-quarters of the population faced the prospect of turning on the tap and having nothing come out at all. CAPE TOWN IS APPROACHING DROUGHT "DAY ZERO," read a typical headline. Climate change, for these kids, was not something to read about in books or to fear off in the distance. It was as present and urgent as thirst itself.

The same was true at the climate strike on the Pacific island nation of Vanuatu, where residents live in fear of further coastal erosion. Their Pacific neighbor, the Solomon Islands, has already lost five small islands to rising water, with six more at severe risk of disappearing forever.

"Raise your voice, not the sea level!" the students chanted.

In New York City, ten thousand kids from dozens of schools found one another in Columbus Circle and proceeded to march to

Trump Tower, chanting "Money won't matter when we're dead!" The older teens in the crowd had vivid memories of when Superstorm Sandy slammed into their coastal city in 2012. "My house got flooded and I was so confused," recalled Sandra Rogers. "And it really made me look into it because you don't learn these things in school."

New York City's huge Puerto Rican community was also out in force on that unseasonably warm day. Some kids arrived draped in the island's flag, a reminder of the relatives and friends still suffering in the aftermath of Hurricane Maria, the 2017 storm that knocked out electricity and water in large parts of the territory for the better part of a year, a total infrastructure breakdown that took the lives of roughly three thousand people.

The mood was fierce, too, in San Francisco, when more than a thousand student strikers shared stories of living with chronic asthma because of polluting industries in their neighborhoods—and then getting a whole lot sicker when wildfire smoke choked the Bay Area just a few months before the strike. The testimonies were similar at walk-outs all over the Pacific Northwest, where smoke from record-breaking fires had blotted out the sun for two summers running. Across the northern border in Vancouver, young people had recently succeeded in pressuring their city council to declare a "climate emergency."

Seven thousand miles away, in Delhi, student strikers braved the ever-present air pollution (often the worst in the world) to shout through white medical masks, "You sold our future, just for profit!" In interviews, some spoke of the devastating floods in Kerala that killed more than four hundred people in 2018.

Australia's coal-addled resource minister declared that "The best thing you'll learn about going to a protest is how to join the dole queue." Undeterred, 150,000 young people poured into plazas in Sydney, Melbourne, Brisbane, Adelaide, and other cities.

This generation of Australians has decided it simply cannot pretend that everything is normal. Not when, at the start of 2019, the South Australian city of Port Augusta had reached an oven-worthy 121°F (49.5°C). Not when half the Great Barrier Reef, the world's largest natural structure made up of living creatures, had turned into a rotting underwater mass grave. Not when, in the weeks leading up to the strike itself, they had seen bushfires combine into a massive blaze in the state of Victoria, forcing thousands to flee their homes, while in Tasmania, wildfires destroyed old-growth rain forests that are unlike any ecosystem in the world. Not when, in January 2019, a combination of extreme temperature swings and poor water management led the entire country to wake up to apocalyptic images of the Darling River clogged with the floating carcasses of one million dead fish.

"You have failed us all so terribly," said fifteen-year-old strike organizer Nosrat Fareha, addressing the political class as a whole. "We deserve better. Young people can't even vote but will have to live with the consequences of your inaction."

There was no student strike in Mozambique; on March 15, the day of the global walkouts, the whole country was bracing for the impact of Cyclone Idai, one of the worst storms in African history, which drove people to take refuge at the tops of trees as the waters rose and would eventually kill more than one thousand people. And then, just six weeks later, while it was still clearing the rubble, Mozambique would be hit by Cyclone Kenneth, yet another record-breaking storm.

Wherever in the world they live, this generation has something in common: they are the first for whom climate disruption on a planetary scale is not a future threat, but a lived reality. And not in a few unlucky hot spots, but on every single continent, with pretty much everything unraveling significantly faster than most scientific models had predicted.

Oceans are warming 40 percent faster than the United Nations predicted just five years ago. And a sweeping study on the state of the Arctic published in April 2019 in *Environmental Research Letters*, led by renowned glaciologist Jason Box, found that ice in various forms is melting so rapidly that the "Arctic biophysical system is now clearly trending away from its 20th Century state and into an unprecedented state, with implications not only within but also beyond the Arctic." In May 2019, the United Nations' Intergovernmental Science-Policy Platform on Biodiversity and Ecosystem Services published a report about the startling loss of wildlife around the world, warning that a million species of animals and plants are at risk of extinction. "The health of ecosystems on which we and all other species depend is deteriorating more rapidly than ever," said the Platform's Chair, Robert Watson. "We are eroding the very foundations of economies, livelihoods, food security, health and quality of life worldwide. We have lost time. We must act now."

And so, just as US schoolchildren now grow up practicing "active shooter drills" starting in kindergarten, many of these students have had school days cancelled because of wildfire smoke, or learned to pack an evacuation bag ahead of hurricanes. A great many children have been forced to leave their homes for good because prolonged drought destroyed their parents' livelihood in Guatemala, or contributed to the outbreak of civil war in Syria.

It has been over three decades since governments and scientists started officially meeting to discuss the need to lower greenhouse gas emissions to avoid the dangers of climate breakdown. In the intervening years, we have heard countless appeals for action that involve "the children," "the grandchildren," and "generations to come." We were told that we owed it to them to move swiftly and embrace change. We were warned that we were failing in our most sacred duty to protect them. It was predicted that they would judge us harshly if we failed to act on their behalf.

Well, none of those emotional pleas proved at all persuasive, at least not to the politicians and their corporate underwriters who could have taken bold action to stop the climate disruption we are all living through today. Instead, since those government meetings began in 1988, global CO_2 emissions have risen by well over 40 percent, and they continue to rise. The planet has warmed by about 1°C since we began burning coal on an industrial scale and average temperatures are on track to rise by as much as four times that amount before the century is up; the last time there was this much carbon dioxide in the atmosphere, humans didn't exist.

As for those children and grandchildren and generations to come who were invoked so promiscuously? They are no longer mere rhetorical devices. They are now speaking (and screaming, and striking) for themselves. And they are speaking up for one another as part of an emerging international movement of children and a global web of creation that includes all those amazing animals and natural wonders that they fell in love with so effortlessly, only to discover that it was all slipping away.

And yes, as foretold, these children are ready to deliver their moral verdict on the people and institutions who knew all about the dangerous, depleted world they would inherit and yet chose not to act.

They know what they think of Donald Trump in the United States and Jair Bolsonaro in Brazil and Scott Morrison in Australia and all the other leaders who torch the planet with defiant glee while denying science so basic that these kids could grasp it easily at age eight. Their verdict is just as damning, if not more so, for the leaders who deliver passionate and moving speeches about the imperative to respect the Paris Climate Agreement and "make the planet great again" (France's Emanuel Macron, Canada's Justin Trudeau, and so many others), but who then shower subsidies, handouts, and licenses on the fossil fuel and agribusiness giants driving ecological breakdown.

Young people around the world are cracking open the heart of the climate crisis, speaking of a deep longing for a future they thought they had but that is disappearing with each day that adults fail to act on the reality that we are in an emergency.

This is the power of the youth climate movement. Unlike so many adults in positions of authority, they have not yet been trained to mask the unfathomable stakes of our moment in the language of bureaucracy and overcomplexity. They understand that they are fighting for the fundamental right to live full lives—lives in which they are not, as thirteen-year-old climate striker Alexandria Villaseñor puts it, "running from disasters."

On that day in March 2019, organizers estimate, there were nearly 2,100 youth climate strikes in 125 countries, with 1.6 million young people participating. That's quite an achievement for a movement that began just eight months earlier with a single fifteen-year-old girl in Stockholm, Sweden.

GRETA'S "SUPERPOWER"

The girl in question is Greta Thunberg, and her story has important lessons about what it will take to protect the possibility of a livable future—and not for some abstract idea of "future generations" but for billions of people alive today.

Like many of her peers, Greta started learning about climate change when she was around eight years old. She read books and watched documentaries about species collapse and melting glaciers. She became obsessed. She learned that burning fossil fuels and eating a meat-based diet were major contributors to planetary destabilization. She discovered that there was a delay between our actions and the planet's reactions, which means that more warming is already locked in, no matter what we do.

As she grew up and learned more, she focused on the scientific predictions about how radically the earth is on track to change by 2040, 2060, and 2080 if we stay on our current course. She made mental calculations about what this would mean to her own life: the shocks she would have to endure, the death that could surround her, the other life forms that would disappear forever, the horrors and privations that would await her own children should she decide to become a parent.

Greta also learned from climate scientists that the worst of this was not a foregone conclusion: that if we took radical action now, reducing emissions by 15 percent a year in wealthy countries like Sweden, then it would dramatically increase the chances of a safe future for her generation and the ones that followed. We could still save some of the glaciers. We could still protect many island nations. We might still avoid massive crop failure that would force hundreds of millions, if not billions, of people to flee their homes.

If all this were true, she reasoned, then "we wouldn't be talking about anything else . . . If burning fossil fuels was so bad that it threatened our very existence, how could we just continue like before? Why were there no restrictions? Why wasn't it made illegal?"

It made no sense. Surely governments, especially in countries with resources to spare, should be leading the charge to achieve a rapid transition within a decade, so that by the time she was in her mid-twenties, consumption patterns and physical infrastructure would be fundamentally transformed.

And yet her government, a self-styled climate leader, was moving much more slowly than that, and indeed, global emissions were continuing to rise. It was madness: the world was on fire, and yet everywhere Greta looked, people were gossiping about celebrities, taking pictures of themselves imitating celebrities, buying new cars and new clothes they didn't need—as if they had all the time in the world to douse the flames.

By age eleven, she had fallen into a deep depression. There were many contributing factors, some related to being different in a school system that expects all kids to be pretty much the same. ("I was the invisible girl in the back.") But there was also a feeling of great sorrow and helplessness about the fast deteriorating state of the planet—and the inexplicable failure of those in power to do much of anything about it.

Thunberg stopped speaking and eating. She became very ill. Eventually, she was diagnosed with selective mutism, obsessive-compulsive disorder, and a form of autism that used to be called Asperger's syndrome. That last diagnosis helped explain why Greta took what she was learning about climate change so much harder and more personally than many of her peers.

People with autism tend to be extremely literal and, as a result, often have trouble coping with cognitive dissonance, those gaps between what we know intellectually and what we do that are so pervasive in modern life. Many people on the autism spectrum are also less prone to imitating the social behaviors of the people around them—they often don't even notice them—and instead tend to forge their own unique path. This often involves focusing with great intensity on areas of particular interest, and frequently having difficulty putting those areas of interest aside (also known as compartmentalization). "For those of us who are on the spectrum," Thunberg says, "almost everything is black or white. We aren't very good at lying, and we usually don't enjoy participating in this social game that the rest of you seem so fond of."

These traits explain why some people with Greta's diagnosis become accomplished scientists and classical musicians, applying their super focus to great effect. It also helps explain why, when Thunberg trained her laser-like attention on climate breakdown, she was completely overwhelmed, with no way to protect herself from the fear and grief. She saw and felt the full implications of the crisis and could not be distracted from it. What's more, the fact that other people in her life (classmates, parents, teachers) seemed relatively unconcerned did not send her reassuring social signals that the situation wasn't really so bad, as such signals do for children who are more socially connected. The apparent lack of concern of those around her terrified Thunberg even more.

To hear Greta and her parents tell it, a big part of emerging from her dangerous depression was finding ways to reduce the unbearable cognitive dissonance between what she had learned about the planetary crisis and how she and her family were living their lives. She convinced her parents to join her in becoming vegan, or at least vegetarian, and, biggest of all, to stop flying. (Her mother is a well-known opera singer, so this was no small sacrifice.)

The amount of carbon kept out of the atmosphere as a result of these lifestyle changes was minute. Greta was well aware of that, but persuading her family to live in a way that began to reflect the planetary emergency helped ease some of the psychic strain. At least now, in their own small ways, they were not pretending that everything was fine.

The most important change Thunberg made, however, had nothing to do with eating and flying. It had to do with finding a way to show the rest of the world that it was time to stop acting like everything was normal when normal would lead straight to catastrophe. If she desperately wanted powerful politicians to put themselves on emergency footing to fight climate change, then she needed to reflect that state of emergency in her own life.

That's how, at age fifteen, she decided to stop doing the one thing all kids are supposed to do when everything is normal: go to school to prepare for their futures as adults.

"Why," Greta wondered, "should we be studying for a future that soon may be no more when no one is doing anything whatsoever to save that future? And what is the point of learning facts in the school system when the most important facts given by the finest science of that same school system clearly means nothing to our politicians and our society."

So, in August 2018, at the start of the school year, Thunberg didn't go to class. She went to Sweden's parliament and camped outside with a handmade sign that read simply, SCHOOL STRIKE FOR THE CLIMATE. She returned every Friday, spending all day there. At first, Greta, in her thrift shop blue hoodie and tousled brown braids, was utterly ignored, like an inconvenient panhandler tugging at the conscience of stressed-out, harried people.

Gradually, her quixotic protest got a bit of press attention, and other students, and a few adults, started visiting with signs of their own. Next came the speaking invitations—first at climate rallies, then at UN climate conferences, at the European Union, at TEDxStockholm, at the Vatican, at the British Parliament. She was even invited to go up that famous mountain in Switzerland to address the rich and mighty at the annual World Economic Summit in Davos.

Every time she spoke, Greta's interventions were short, unadorned, and utterly scathing. "You are not mature enough to tell it like it is," she told the climate change negotiators in Katowice, Poland. "Even that burden you leave to us children." To British MPs she asked, "Is my English OK? Is the microphone on? Because I'm beginning to wonder."

To the rich and mighty at Davos who praised her for giving them hope, she replied, "I don't want your hope . . . I want you to panic. I want you to feel the fear I feel every day. I want you to act. I want you to act as you would in a crisis. I want you to act as if the house is on fire, because it is."

To those in the rarified crowd of CEOs, celebrities, and politicians who spoke of climate disruption as if it were a problem of universal human shortsightedness, she shot back, "If everyone is guilty, then no one is to blame, and someone *is* to blame . . . Some people, some companies, some decision-makers in particular know exactly what priceless values they have been sacrificing to continue making unimaginable amounts of money." She paused, took a breath, and said, "And I think many of you here today belong to that group of people."

Her sharpest rebuke to the Davos set was wordless. Rather than staying in one of the five-star hotel rooms on offer, she braved -18°C temperatures (0°F) to sleep outside in a tent, snuggled in a bright yellow sleeping bag. ("I'm not a great fan of heat," Greta told me.)

When she spoke to these rooms filled with adults in suits, who clapped and filmed her on their smartphones as if she were a novelty act, Thunberg's voice rarely trembled. But the depth of her feeling—of loss, of fear, of love for the natural world—was always unmistakable. "I beg you," Thunberg said in an emotional address to members of the European Parliament in April 2019. "Please do not fail on this."

Even if the speeches did not dramatically change the actions of the policymakers in those stately rooms, they did change the actions of a great many people outside them. Nearly every video of the fiery-eyed girl went viral. It was as if by yelling "Fire!" on our crowded planet, she had given countless others the confidence they needed to believe their own senses and smell the smoke drifting in under all those tightly closed doors.

It was more than that, too. Listening to Thunberg speak about how our collective climate inaction had nearly stolen her will to live seemed to help others feel the fire of survival in their own bellies. The clarity of Greta's voice gave validation to the raw terror so many of us have been suppressing and compartmentalizing about what it means to be alive amid the sixth great extinction and surrounded by scientific warnings that we are flat out of time.

Suddenly, children around the world were taking their cues from Greta, the girl who takes social cues from no one, and were organizing student strikes of their own. At their marches, many held up placards quoting some of her most piercing words: I WANT YOU TO PANIC, OUR HOUSE IS ON FIRE. At a massive school strike in Düsseldorf, Germany, demonstrators held aloft a giant papier-mâché puppet of Greta, brow furrowed and braids dangling, like the patron saint of pissed-off kids everywhere.

Greta's voyage from invisible schoolgirl to global voice of conscience is an extraordinary one, and looked at more closely, it has a lot to teach about what it is going to take for all of us to get to safety. Thunberg's overarching demand is for humanity as a whole to do what she did in her own family and life: close the gap between what we know about the urgency of the climate crisis and how we behave. The first stage is to name the emergency, because only once we are on emergency footing will we find the capacity to do what is required.

In a way, she is asking those of us whose mental wiring is more typical—less prone to extraordinary focus and more capable of living with moral contradictions—to be more like her. She has a point.

During normal, nonemergency times, the capacity of the human mind to rationalize, to compartmentalize, and to be distracted easily is an important coping mechanism. All three of these mental tricks help us get through the day. It's also extremely helpful to look unconsciously to our peers and role models to figure out how to feel and act—those social cues are how we form friendships and build cohesive communities.

When it comes to rising to the reality of climate breakdown, however, these traits are proving to be our collective undoing. They are reassuring us when we should not be reassured. They are distracting us when we should not be distracted. And they are easing our consciences when our consciences should not be eased.

In part, this is because if we were to decide to take climate disruption seriously, pretty much every aspect of our economy would have to change, and there are many powerful interests that like things as they are. Not least the fossil fuel corporations, which have funded a decades-long campaign of disinformation, obfuscation, and straight-up lies about the reality of global warming.

As a result, when most of us look around for social confirmation of what our hearts and heads are telling us about climate disruption, we are confronted with all kinds of contradictory signals, telling us instead not to worry, that it's an exaggeration, that there are countless more important problems, countless shinier objects to focus on, that we'll never make a difference anyway, and so on. And it most certainly doesn't help that we are trying to navigate this civilizational crisis at a moment when some of the most brilliant minds of our time are devoting vast energies to figuring out ever-more-ingenious tools to keep us running around in digital circles in search of the next dopamine hit.

This may explain the odd space that the climate crisis occupies in the public imagination, even among those of us who are actively terrified of climate collapse. One minute we're sharing articles about the insect apocalypse and viral videos of walruses falling off cliffs because sea ice loss has destroyed their habitat, and the next, we're online shopping and willfully turning our minds into Swiss cheese by scrolling through Twitter or Instagram. Or else we're binge-watching Netflix shows about the zombie apocalypse that turn our terrors into entertainment, while tacitly confirming that the future ends in collapse anyway, so why bother trying to stop the inevitable? It also might explain the way serious people can simultaneously grasp how close we are to an irreversible tipping point and still regard the only people who are calling for this to be treated as an emergency as unserious and unrealistic.

"I think in many ways that we autistic are the normal ones, and the rest of the people are pretty strange," Thunberg has said, adding that it helps not to be easily distracted or reassured by rationalizations. "Because if the emissions have to stop, then we must stop the emissions. To me that is black or white. There are no gray areas when it comes to survival. Either we go on as a civilization or we don't. We have to change." Living with autism is anything but easy—for most people, it "is an endless fight against schools, workplaces and bullies. But under the right circumstances, given the right adjustments it *can* be a superpower."

The wave of youth mobilization that burst onto the scene in March 2019 is not the result of one girl and her unique way of seeing the world. Greta is quick to note that she was inspired by another group of teenagers who rose up against a different kind of failure to protect their futures: the students in Parkland, Florida, who led a national wave of class walkouts demanding tough controls on gun ownership after seventeen people were murdered at their school in February 2018.

Nor is Thunberg the first person with tremendous moral

clarity to yell "Fire!" in the face of the climate crisis. It has happened multiple times over the past several decades; indeed, it is something of a ritual at the annual UN summits on climate change. But perhaps because these earlier voices belonged to brown and black people from the Philippines, the Marshall Islands, and South Sudan, those clarion calls were one-day stories, if that. Thunberg is also quick to point out that the climate strikes themselves were the work of thousands of diverse student leaders, their teachers, and supporting organizations, many of whom had been raising the climate alarm for years.

As a manifesto put out by British climate strikers put it, "Greta Thunberg may have been the spark, but we're the wildfire."

For a decade and a half, ever since reporting from New Orleans with water up to my waist after Hurricane Katrina, I have been trying to figure out what is interfering with humanity's basic survival instinct—why so many of us aren't acting like our house is on fire when it so clearly is. I have written books, made films, delivered countless talks and cofounded an organization (The Leap) devoted, in one way or another, to exploring this question and trying to help align our collective response to the scale of the climate crisis.

It was clear to me from the start that the dominant theories about how we had landed on this knife edge were entirely insufficient. We were failing to act, it was said, because politicians were trapped in short-term electoral cycles, or because climate change seemed too far off, or because stopping it was too expensive, or because the clean technologies weren't there yet. There was some truth in all the explanations, but they were also becoming less true over time. The crisis wasn't far off; it was banging down our doors. The price of solar panels has plummeted and now rivals that of fossil fuels. Clean tech and renewables create far more jobs than coal, oil, and gas. As for the supposedly prohibitive costs, trillions have been marshaled for endless wars, bank bailouts, and subsidies for fossil fuels, in the same years that coffers have been virtually empty for climate transition. There had to be more to it.

This book, made up of long-form reporting, think pieces, and public talks written over the span of a decade, tracks my own attempt to probe a different set of barriers—some economic, some ideological, but others related to the deep stories about the right of certain people to dominate land and the people living closest to it, stories that underpin Western culture. The essays here return frequently to the kinds of responses that might succeed in toppling those narratives, ideologies, and economic interests, responses that weave seemingly disparate crises (economic, social, ecological, and democratic) into a common story of civilizational transformation. Today, that kind of bold vision increasingly goes under the banner of a "Green New Deal."

I have chosen to organize the pieces in the order in which they were written, with the original month and year appearing at the start. It's a structure that, while involving the occasional return to a theme, reflects the evolution of my own analysis as I tested these ideas out in the world and worked in collaboration with countless friends and colleagues in the global climate justice movement. With the exception of the final essays specifically about the Green New Deal, which have been expanded significantly, I resisted the urge to modify the texts and instead left them pretty much intact, clarifying time frames and adding updates in footnotes and postscripts here and there.

Keeping the pieces in chronological order has one main benefit. It's a nagging reminder that we are in a fast-moving crisis, even though it may not always seem so. In the short decade spanned by this book, the planet has undergone enormous and irreparable damage, from rapid disappearance of Arctic sea ice to mass die-offs of coral reefs. The part of the world where my family is from, the west coast of British Columbia, has seen the collapse of certain species of Pacific salmon that hold entire magnificent ecosystems on their backs.

The political map has also changed dramatically over this decade. It has seen the resurgence of an increasingly violent hard right, a force that is gaining power across the globe by stoking hatred against ethnic, religious, and racial minorities, often manifesting as xenophobic hatred directed at the growing numbers of people who have been forced to leave their homelands. These planetary and political trends are, I am convinced, in a kind of lethal dialogue with one another.

For me, the temporal references throughout the book are like the hourglass on that student striker's sign: relentless evidence that as our societies fail to act like our house is on fire, the house does not just sit there burning like some sort of fixed, looping GIF. The conflagration gathers more and more heat—and irreplaceable parts of the house actually do burn to the ground. Gone, forever.

My primary emphasis in this book is on the countries sometimes referred to as the Anglosphere (the United States, Canada, Australia, and the United Kingdom) and on some non-English-speaking parts of Europe. In part, this is by happenstance—I currently live and work in the United States, have spent most of my life in Canada, and have participated extensively in climate change debates and initiatives in Australia, the United Kingdom, and other parts of western Europe.

This focus stems mainly, however, from my ongoing attempt to understand why the governments of these countries have proven particularly belligerent when it comes to meaningful climate action.

There is still a significant (though thankfully shrinking) segment of the population in each of these nations that chooses to deny the basic fact that human activity is causing the planet to dangerously warm, a glaring truth that is uncontroversial and uncontested in most parts of the world.

Even when outright denial recedes and a more progressive environmental era seems to dawn (in the United States under Barack Obama, in Canada under Justin Trudeau), it is still extremely difficult for these governments to accept the overwhelming scientific evidence that we need to stop extending the fossil fuel frontier and, in fact, need to start winding down existing production. Australia, despite its wealth, insists on massively expanding coal production in the teeth of the climate crisis; Canada has done the same with the Alberta tar sands; the United States has done the same with Bakken oil, fracked gas, and deepwater drilling, becoming the world's largest oil exporter; the United Kingdom has attempted to ram through fracking operations despite fierce opposition and evidence linking it to earthquakes.

In trying to make sense of this, I explore some of the specific ways that these nations led the way in forging the global supply chain that gave birth to modern capitalism, the economic system of limitless consumption and ecological depletion at the heart of the climate crisis. It's a story that begins with people stolen from Africa and lands stolen from Indigenous peoples, two practices of brutal expropriation that were so dizzyingly profitable that they generated the excess capital and power to launch the age of fossil fuel–led industrial revolution and, with it, the beginning of human-driven climate change. It was a process that required, from the start, pseudoscientific as well as theological theories of white and Christian supremacy, which is why the late political theorist Cedric Robinson argued that the economic system birthed by the convergence of these fires should more aptly be called "racial capitalism."

Alongside the theories that rationalized treating humans as raw capitalist assets to exhaust and abuse without limit were theories that justified treating the natural world (forests, rivers, land and water animals) in precisely the same way. Millennia of accumulated human wisdom about how to safeguard and regenerate everything from forests to fish runs were swept away in favor of a new idea that there was no limit to humanity's ability to control the natural world, nor to how much wealth could be extracted from it without fear of consequence.

These ideas about nature's boundlessness are not incidental to the nations of the Anglosphere; they are foundational myths, woven deep into national narratives. The huge natural wealth of the lands that would become the United States, Canada, and Australia were, from their very first contact with European ships, imagined as sort of body-double nations for colonial powers that were running out of nature to exhaust back home. No more. With the "discovery" of these seemingly limitless "new worlds," God had granted a reprieve: *New* England, *New* France, *New* Amsterdam, *New* South Wales—proof positive that Europeans would never run out of nature to exhaust. And when one swath of this new territory grew depleted or crowded, the frontier would simply advance and *new* "new worlds" would be named and claimed.

In these pages, I explore this original imaginative sin as it relates to the climate crisis from many different vantage points: the black death of BP's oil spreading through the Gulf of Mexico; the Vatican under Pope Francis's "ecological conversion"; Trump's grab-and-go America; the die-off in the Great Barrier Reef, where Captain James Cook's ship (a converted coal barge) once ran aground; and more. I also try to understand the intersection of these collapsing mythologies, as nature reveals itself to be anything but infinitely exhaustible and abusable, and the terrifying resurgence of the ugliest and most violent parts of these colonial narratives throughout the Anglosphere—the parts about the right of supposedly superior white Christians to inflict tremendous violence on those they have decided to classify as beneath them in a brutal hierarchy of humanity.

I am not arguing that these nations are the sole drivers of our ecological breakdown, not by any means. Our crisis is global, and many other countries have polluted recklessly during this same period. (Pick your petrostate, or watch China's and India's emissions soar.) But rapid acceleration of climate breakdown has occurred simultaneous to, and as a direct result of, the successful globalization of the high consumer lifestyle birthed in the nations I write about in this book. These are, moreover, the nations that have been polluting at extremely high levels for centuries and that, therefore, had an obligation, under the UN Framework Convention on Climate Change that their governments all signed, to lead the way on emission reduction before the developing world. As US officials used to say during the 2003 invasion of Iraq, "We broke it, we bought it."

A PEOPLE'S EMERGENCY

Yet, as deep as our crisis runs, something equally deep is also shifting, and with a speed that startles me. As I write these words, it is not only our planet that is on fire. So are social movements rising up to declare, from below, a people's emergency. In addition to the wildfire of student strikes, we have seen the rise of Extinction Rebellion, which exploded on the scene and kicked off a wave of nonviolent direct action and civil disobedience, including a mass shutdown of large parts of Central London. Extinction Rebellion is calling on governments to treat climate change as an emergency, to rapidly transition to 100 percent renewable energy in line with climate science, and to democratically develop the plan for how to implement that transition through citizens' assemblies. Within days of its most dramatic actions in April 2019, Wales and Scotland both declared a state of "climate emergency," and the British Parliament, under pressure from opposition parties, quickly followed suit.

In this same period in the United States, we have seen the meteoric rise of the Sunrise Movement, which burst onto the political stage when it occupied the office of Nancy Pelosi, the most powerful Democrat in Washington, DC, one week after her party had won back the House of Representatives in the 2018 midterm elections. Wasting no time on congratulations, the Sunrisers accused the party of having no plan to respond to the climate emergency. They called on Congress to immediately adopt a rapid decarbonization framework, one as ambitious in speed and scope as Franklin D. Roosevelt's New Deal, the sweeping package of policies designed to battle the poverty of the Great Depression and the ecological collapse of the Dust Bowl.

As a writer and organizer, I have been part of the global climate movement for years, and it has taken me to many large marches and mass actions, including the four-hundred-thousand-strong People's Climate March in New York City in 2014. I have covered and participated in major UN climate summits that made lofty promises to rise to humanity's existential challenge (Copenhagen in 2009, Paris in 2014). As a board member of the climate campaign group 350.org, I was part of kick-starting the fossil fuel divestment movement, which, as of December 2018, succeeded in getting $8 trillion in investment wealth to commit to selling off its holdings in fossil fuel companies. And I have been part of several movements, some of them successful, to stop the laying of new oil pipelines.

The activism we are seeing today builds on this history and also changes the equation completely. Though many of the efforts just described were large, they still engaged primarily with self-identified environmentalists and climate activists. If they did reach beyond those circles, the engagement was rarely sustained for more than a single march or pipeline fight. Outside the climate movement, there was still a way that the planetary crisis could be forgotten for months on end or go barely mentioned during pivotal election campaigns.

Our current moment is markedly different, and the reason for that is twofold: one part having to do with a mounting sense of peril, the other with a new and unfamiliar sense of promise.

THE RADICALIZING POWER OF CLIMATE SCIENCE

One month before the Sunrisers occupied the office of soon-to-be House Speaker Nancy Pelosi, the UN Intergovernmental Panel on Climate Change (IPCC) published a report that had a greater impact than any publication in the thirty-one-year history of the Nobel Peace Prize–winning organization.

The report examined the implications of keeping the increase in planetary warming below 1.5 degrees Celsius (2.7°F). Given the worsening disasters we are already seeing with about 1°C of warming, it found that keeping temperatures below the 1.5°C threshold is humanity's best chance of avoiding truly catastrophic unraveling.

But doing that would be extremely difficult. According to the UN World Meteorological Organization, we are on a path to warming the world by 3–5°C by the end of the century. Turning our economic ship around in time to keep the warming below 1.5°C would require, the IPCC authors found, cutting global emissions approximately in half in a mere twelve years—that's eleven years as this book goes to press—and getting to net-zero carbon emissions by 2050. Not just in one country but in every major economy. And because carbon dioxide in the atmosphere has already dramatically surpassed safe levels, it would also require drawing a great deal of that down, whether through unproven and expensive carbon capture technologies or the old-fashioned ways: by planting billions of trees and other carbon-sequestering vegetation.

Pulling off this high-speed pollution phaseout, the report establishes, is not possible with singular technocratic approaches like carbon taxes, though those tools must play a part. Rather, it requires deliberately and immediately changing how our societies produce energy, how we grow our food, how we move ourselves around, and how our buildings are constructed. What is needed, the report's summary states in its first sentence, is "rapid, far-reaching and unprecedented changes in all aspects of society."

This was not the first terrifying climate report by any means, nor the first unequivocal call from respected scientists for radical emission reduction. My bookshelves are crowded with these findings. But like Greta Thunberg's speeches, the starkness of the IPCC's call for root-and-branch societal change, and the shortness of the time line it laid out for pulling it off, focused the public mind like nothing before.

A big part of that has to do with the source. After governments came together to recognize the threat of global warming in 1988, the United Nations created the IPCC to provide policymakers with the most reliable information possible to inform their decisions. For this reason, the panel synthesizes all the best science to come up with projections that a great many scientists need to agree on before anything is made public—and even then, nothing can go out before the governments themselves sign off.

Because of this laborious process, IPCC projections have been notoriously conservative, often dangerously underestimating risk. And yet here was a report, drawing on some six thousand sources, created by nearly one hundred authors and review editors, saying in no uncertain terms that if governments did as little to cut emissions as they were currently pledging to do, we were headed toward consequences including sea level rise that would swallow coastal cities, the total die-off of coral reefs, and droughts that would wipe out crops in huge parts of the globe.

Today's high school students will still be in their twenties when global emissions need to have already been cut in half to avoid those outcomes. And yet the fateful decisions about whether those cuts will happen—decisions that will shape their entire lives—are being made well before most of them even have the right to vote.

It was against this backdrop that 2019's cascade of large and militant climate mobilizations unfolded. Again and again at the strikes and protests, we heard the words "We have only twelve years." Thanks to the IPCC's unequivocal clarity, as well as direct and repeated experience with unprecedented weather, our conceptualization of this crisis is shifting. Many more people are beginning to grasp that the fight is not for some abstraction called "the earth." We are fighting for our lives. And we don't have twelve years anymore; now we have only eleven. And soon it will be just ten.

ENTER THE GREEN NEW DEAL

As powerful a motivator as the IPCC report has proven to be, perhaps an even more important factor has to do with this book's subtitle: the calls coming from many different quarters in the United

States and around the world for governments to respond to the climate crisis with a sweeping Green New Deal. The idea is a simple one: in the process of transforming the infrastructure of our societies at the speed and scale that scientists have called for, humanity has a once-in-a-century chance to fix an economic model that is failing the majority of people on multiple fronts. Because the factors that are destroying our planet are also destroying people's quality of life in many other ways, from wage stagnation to gaping inequalities to crumbling services to the breakdown of any semblance of social cohesion. Challenging these underlying forces is an opportunity to solve several interlocking crises at once.

In tackling the climate crisis, we can create hundreds of millions of good jobs around the world, invest in the most systematically excluded communities and nations, guarantee health care and child care, and much more. The result of these transformations would be economies built both to protect and to regenerate the planet's life support systems and to respect and sustain the people who depend on them. It would also strive for something more amorphous but equally important: at a time when we find ourselves increasingly divided into hermetically sealed information bubbles, with almost no shared assumptions about what we can trust or even about what is real, a Green New Deal could instill a sense of collective, higher purpose—a set of concrete goals that we are all working toward together. In scale if not specifics, the Green New Deal proposal takes its inspiration from Franklin Delano Roosevelt's original New Deal, which responded to the misery and breakdown of the Great Depression with a flurry of policies and public investments, from introducing Social Security and minimum wage laws, to breaking up the banks, to electrifying rural America and building a wave of low-cost housing in cities, to planting more than two billion trees and launching soil protection programs in regions ravaged by the Dust Bowl.

The various plans that have emerged for a Green New Deal–style transformation envision a future where the difficult work of transition has been embraced, including sacrifices in profligate consumption. But in exchange, day-to-day life for working people has been improved in countless ways, with more time for leisure and art, truly accessible and affordable public transit and housing, yawning racial and gender wealth gaps closed at last, and city life that is not an unending battle against traffic, noise, and pollution.

Long before the IPCC's 1.5°C report, the climate movement had focused on the perilous future we faced if politicians failed to act. We popularized and shared the latest terrifying science. We said no to new oil pipelines, gas fields, and coal mines; no to universities, local governments, and unions investing endowments and pensions in the companies behind these projects; no to politicians who denied climate change and no to politicians who said all the right things but did the wrong ones. All this was critical work, and it remains so. But while we raised the alarm, only the relatively small "climate justice" wing of the movement focused its attention on the kind of economy and society we wanted instead.

That was the game changer of the Green New Deal bursting into the political debate in November 2018. Wearing shirts that read WE HAVE A RIGHT TO GOOD JOBS AND A LIVABLE FUTURE, hundreds of young members of the Sunrise Movement chanted for a Green New Deal as they lined the halls of Congress shortly after the 2018 midterms. There was finally a big and bold "yes" to pair with the climate movement's many "no's," a story of what the world could look like after we embraced deep transformation, and a plan for how to get there.

The Green New Deal's roots-up approach to the climate crisis is not itself new. This kind of "climate justice" framework (as opposed to the more generic "climate action") has been attempted locally for many years, with its origins in the Latin American and US environmental justice movements. And the concept of a Green New Deal has made it into the platforms of a few small Green parties around the world.

My 2014 book, *This Changes Everything: Capitalism vs. the Climate*, explored this kind of holistic approach in depth. The historical precedent I used back then came from a Bolivian climate negotiator named Angélica Navarro Llanos, who delivered a blistering address to a 2009 UN climate summit: "We need a massive mobilization larger than any in history. We need a Marshall Plan for the Earth," she declared, invoking the ways that the United States, fearing an ascendant Soviet Union, had helped rebuild large parts of Europe after World War II. "This plan must mobilize financing and technology transfer on scales never seen before. It must get technology onto the ground in every country to ensure we reduce emissions while raising people's quality of life. We have only a decade."

We wasted the entire decade following that call with tinkering and denial, and we will never get back the wonders that are gone as a result—or the lives and livelihoods destroyed because of it. Navarro Llanos and her fellow Bolivians have watched the majestic glaciers that provide fresh water for the metropolitan area of La Paz (home to 2.3 million people) recede with alarming speed. In 2017, reservoirs ran so low that water rationing was introduced for the first time in the capitol and a state of emergency had to be declared across the country.

But that lost decade does not make Navarro Llanos's prescient call less relevant—it makes it far more so, given that, as the IPCC report made so clear, hundreds of millions of lives hang in the balance with every half degree of warming we either enable or avoid.

• • •

Something else has changed since that call was issued a decade ago. Before, when social movements and small country governments made these demands, it felt as if we were screaming into a political void. There was really no cohort in the governments of the wealthiest countries on the planet willing to entertain this kind of emergency approach to the climate crisis. Trickle-down market mechanisms were the only ones on offer. And when there was an economic downturn, even those inadequate offerings evaporated.

That is no longer the case today. There is now a bloc of politicians in the United States, Europe, and elsewhere, some just a decade older than the young climate activists in the streets, ready to translate the urgency of the climate crisis into policy, and to connect the dots among the multiple crises of our times. Most prominent among this new political breed is Alexandria Ocasio-Cortez, who, at twenty-nine, became the youngest woman ever elected to the US Congress.

Introducing a Green New Deal was part of the platform she ran on. Quickly after winning the elections, several members of the small group of young congresswomen sometimes referred to as the "squad" pledged their support for the bold initiative, particularly Rashida Tlaib of Detroit and Ayanna Pressley of Boston.

So, when hundreds of members of the Sunrise Movement came to Washington after the midterms to hold demonstrations and sit-ins, these newly elected representatives did not keep a safe distance from the rabble-rousers. Instead, they joined them, with Tlaib speaking at one of their rallies (and bringing candy for the crowd to help keep their energy up) and Ocasio-Cortez dropping by their sit-in at Nancy Pelosi's office.

"I just want to let you all know how proud I am of each and every single one of you for putting yourselves and your bodies and everything on the line to make sure that we save our planet, our generation, and our future," she told the demonstrators, reminding them that "my journey here started at Standing Rock," a reference to her decision to run for Congress after participating in the anti-pipeline protests led by the Standing Rock Sioux.

Then, three months later, Ocasio-Cortez, along with Senator Ed Markey of Massachusetts, stood in front of the Capitol and launched a formal resolution for a Green New Deal, a rough outline of the key planks of the transformation. The Green New Deal resolution begins with the terrifying science and short time lines in the IPCC report and calls for the United States to launch a moon shot approach to decarbonization, attempting to reach net-zero emissions in just one decade, in line with getting the entire world there by mid-century.

As part of this sweeping transition, it calls for huge investments in renewable energy, energy efficiency, and clean transportation. It states that workers moving from high-carbon industries to green ones should have their wage levels and benefits protected, and it guarantees a job to all who want to work. It also calls for the communities who have borne the toxic brunt of dirty industries, so many of them Indigenous, black, and brown, not only to benefit from the transitions but to help design them at the local level. And as if all this weren't enough, it folds in key demands from the growing Democratic Socialist wing of the Democratic Party: free universal health care, child care, and higher education.

By previous standards, the framework was shockingly bold and progressive, but there was so much momentum for it, particularly among young voters, that in short order it became a litmus test for large parts of the party. By May 2019, with the race to head the Democratic Party in full swing, the majority of leading presidential hopefuls claimed to support it, including Bernie Sanders, Elizabeth Warren, Kamala Harris, Cory Booker, and Kirsten Gillibrand. It had been endorsed, meanwhile, by 105 members of the House and Senate.

The emergence of the Green New Deal means there is now not only a political framework for meeting the IPCC targets in the United States but also a clear (if long-shot) path to turning that framework into law. The plan is pretty straightforward: elect a strong supporter of the Green New Deal in the Democratic primaries; take the White House, the House, and the Senate in 2020; and start rolling it out on day one of the new administration (the way FDR did with the original New Deal in the famous "first 100 days," when the newly elected president pushed fifteen major bills through Congress).

If the IPCC report was the clanging fire alarm that grabbed the attention of the world, the Green New Deal is the beginning of a fire safety and prevention plan. And not a piecemeal approach that merely trains a water gun on a blazing fire, as we have seen so many times in the past, but a comprehensive and holistic plan to actually put out the fire. Especially if the idea spreads around the world—which is already beginning to happen.

Indeed, in January 2019, the political coalition European Spring (an outgrowth of a project called DiEM25, where I serve on the advisory panel) launched a Green New Deal for Europe, a sweeping and detailed plan to embed an agenda of rapid decarbonization within a broader social and economic justice agenda: "From a green investment programme to drive through the world's ecological transition to clear action on ending the scandal of tax havens; from humane and effective migration policy to a clear plan to tackle poverty in our continent; from a Workers' Compact to a European Convention on Women's Rights and much more, the Green New Deal is the go-to document for anyone wishing to break the dogma of 'There Is No Alternative' and bring back hope to our continent," the coalition announced.

In Canada, a broad coalition of organizations has come together to call for a Green New Deal, with the leader of the New Democratic Party adopting the frame (if not its full ambition) as one of his policy planks. The same is true in the United Kingdom, where the opposition Labour Party is, as I write, in the midst of intense negotiations over whether to adopt a Green New Deal–style platform similar to the one being proposed in the United States.

The various versions of a Green New Deal that have emerged in the past year have something in common. Where previous policies were minor tweaks to incentives that were designed to cause minimal disruption to the system, a Green New Deal approach is a major operating system upgrade, a plan to roll up our sleeves and actually get the job done. Markets play a role in this vision, but markets are not the protagonists of this story—people are. The workers who will build the new infrastructure, the residents who will breathe the clean air, who will live in the new affordable green housing and benefit from the low-cost (or free) public transit.

Those of us who advocate for this kind of transformative platform are sometimes accused of using the climate crisis to advance a socialist or anticapitalist agenda that predates our focus on the climate crisis. My response is a simple one. For my entire adult life, I have been involved in movements confronting the myriad ways that our current economic systems grinds up people's lives and landscapes in the ruthless pursuit of profit. My first book, *No Logo*, published almost exactly twenty years ago, documented the human and ecological costs of corporate globalization, from the sweatshops of Indonesia to the oil fields of the Niger Delta. I have seen teenage girls treated like machines to make our machines, and seen mountains and forests turned to trash heaps to get at the oil, coal, and metals beneath.

The painful, even lethal, impacts of these practices were

impossible to deny; it was simply argued that they were the necessary costs of a system that was creating so much wealth that the benefits would eventually trickle down to improve the lives of nearly everyone on the planet. What has happened instead is that the indifference to life that was expressed in the exploitation of individual workers on factory floors and in the decimation of individual mountains and rivers has instead trickled up to swallow our entire planet, turning fertile lands into salt flats, beautiful islands into rubble, and draining once vibrant reefs of their life and color.

I freely admit that I do not see the climate crisis as separable from the more localized market-generated crises that I have documented over the years; what is different is the scale and scope of the tragedy, with humanity's one and only home now hanging in the balance. I have always had a tremendous sense of urgency about the need to shift to a dramatically more humane economic model. But there is a different quality to that urgency now because it just so happens that we are all alive at the last possible moment when changing course can mean saving lives on a truly unimaginable scale.

None of this means that every climate policy must dismantle capitalism or else it should be dismissed (as some critics have absurdly claimed)—we need every action possible to bring down emissions, and we need them now. But it does mean, as the IPCC has so forcefully confirmed, that we will not get the job done unless we are willing to embrace systemic economic and social change.

HISTORY AS TEACHER—AND WARNING

There are, among emission-reduction experts, long-running debates about which precedents from history to invoke to help inspire the kind of sweeping, economy-wide transformations the climate crisis demands. Many clearly favor FDR's New Deal, because it showed how radically both a society's infrastructure and its governing values can be altered in the span of one decade. And the results are indeed striking. During the New Deal decade, more than 10 million people were directly employed by the government; most of rural America got electricity for the first time; hundreds of thousands of new buildings and structures were built; 2.3 billion trees were planted; 800 new state parks were developed; and hundreds of thousands of public works of art were created.

In addition to the immediate benefits of pulling millions of Depression-ravaged families out of poverty, this period of frenetic public investment left behind a lasting legacy that, despite decades of attempts to dismantle it, survives to this day. Historian Neil Maher, in his book *Nature's New Deal*, provides a helpful snapshot:

> Today, we drive on roads laid out by the Works Progress Administration, drop off our children and pick up books at schools and libraries built by the Public Works Administration, and even drink water flowing from reservoirs constructed by the Tennessee Valley Authority. These and other New Deal Programs . . . dramatically transformed the natural environment. They also altered American politics by introducing the New Deal to the American public in ways that raised popular support for Roosevelt's liberal welfare state.

Others insist that the only precedents that show the scale and speed of change required in the face of the climate crisis are the World War II mobilizations that saw Western powers transforming their manufacturing sectors and consumption patterns to fight Hitler's Germany. It was certainly a dizzying level of change: Factories were retooled to produce ships, planes, and weapons. To free up food and fuel for the military, citizens changed their lifestyles dramatically: in Britain, driving for anything other than necessity virtually ceased; between 1938 and 1944, use of public transit went up by 87 percent in the United States and by 95 percent in Canada. In 1943 in the United States, twenty million households (representing three-fifths of the population) had "Victory gardens" in their yards, growing fresh vegetables that accounted for 42 percent of all those consumed that year.

Some argue that a better analogy than the war effort was the reconstruction afterward—specifically, the Marshall Plan, a kind of New Deal for Western and southern Europe. In West Germany, the US government spent billions to rebuild a mixed economy that would have broad-based support and would undercut the rising support for socialism (while providing a growing market for US exports). That meant direct job creation by the state, huge investment in the public sector, subsidies for German firms, and support for strong labor unions. The effort was widely regarded as Washington's most successful diplomatic initiative.

Each precedent has its own glaring weaknesses and contradictions. The US military alone is, according to the Union of Concerned Scientists, "the largest institutional consumer of oil in the world." And warfare, with its devastating costs to humanity, nature, and democracy, is no model for social change. The climate threat, moreover, will never feel as menacing as Nazis on the march—at least not until it is significantly too late for our behaviors to have a meaningful impact.

The wartime mobilizations, and the huge rebuilding efforts afterward, were certainly ambitious, but they were also highly centralized, top-down transformations. If we defer to central governments in that way in the face of the climate crisis, we should expect highly corrupt measures that further concentrate power and wealth in the hands of a few big players, not to mention systemic attacks on human rights, a phenomenon I have traced repeatedly in my work on disaster capitalism in the aftermath of wars, economic shocks, and extreme weather events. A climate change shock doctrine is a real and present danger, the first signs of which I discuss in these pages.

The New Deal makes a far-from-ideal analogy as well. Most of its programs and protections were designed in a push and pull with social movements, as opposed to merely handed down from on high like the wartime measures. But the New Deal fell short of pulling the US economy out of economic depression, its main goal, and its programs overwhelmingly favored white, male workers. Agricultural and domestic workers (many of them black) were left out, as were many Mexican immigrants (some one million of whom faced deportation in the late 1920s and 1930s), and the Civilian Conservation Corps segregated African American participants and excluded women (except at one camp, where the latter learned canning and other domestic tasks). And while Indigenous peoples won some gains under New Deal programs, land rights were violated by both massive infrastructure projects and some conservation efforts. New Deal relief agencies, particularly in the southern states, were notorious for their biases against unemployed African American and Mexican American families.

The Ocasio-Cortez/Markey Green New Deal resolution goes to considerable lengths to outline how it plans to avoid repeating these injustices, listing as one of its core goals "stopping current, preventing future, and repairing historic oppression of indigenous peoples, communities of color, migrant communities, deindustrialized communities, depopulated rural communities, the poor, low-income workers, women, the elderly, the unhoused, people with disabilities, and youth." As Congresswoman Ayanna Pressley said at a town hall in Boston, "This is not just an opportunity to fix . . . the first New Deal, but also to transform the economy."

The biggest limitation with all these historical comparisons, from the New Deal through to the Marshall Plan, is that, together, they succeeded in kick-starting and massively expanding the high-carbon lifestyle of suburban sprawl and disposable consumption that is at the heart of today's climate crisis. The tough truth, as the IPCC's bombshell report stated explicitly, is that "there is no historical precedent for the scale of the necessary transitions, in particular in a socially and economically sustainable way"—a reference to the fact that global emissions have only ever dropped significantly during times of deep economic crisis, such as the Great Depression and after the collapse of the Soviet Union, and that the wars that spurred rapid fire societal transformations were humanitarian and ecological disasters.

My own view is that as flawed as each historical analogy necessarily is, each is still useful to study and invoke. Every one, in its own way, presents a sharp contrast to how governments have responded to climate breakdown thus far. Over two and a half decades, we have seen the creation of complex carbon markets; the occasional small carbon tax; the replacement of one fossil fuel (coal) with another (gas); various incentives for consumers to buy different kinds of lightbulbs and energy-efficient home appliances; and offers from companies to opt for greener alternatives if we are willing to pay more for them. Yet, only a few countries (most significantly, Germany and China) have made serious enough investments in the renewable sector to see a rollout at anything like the speed required.

We are slowly starting to see a shift to a more aggressive regulatory approach in a handful of countries, invariably as a result of strong social movement pressure. A few countries, states, and provinces have placed bans or moratoriums on fracking for gas. The government of New Zealand, significantly, has announced it will no longer issue leases for offshore oil drilling. Norway's government has announced plans to prohibit the sale of cars with internal combustion engines by 2025, a move that will certainly accelerate a shift to electric vehicles if its aggressive targets spread to other countries. But no national government of a wealthy country has been willing to have a frank discussion about the need for high consumers to consume less or for fossil fuel companies to pay to clean up the mess they created.

And how could it have been otherwise? The past forty years of economic history have been a story of systematically weakening the power of the public sphere, unmaking regulatory bodies, lowering taxes for the wealthy, and selling off essential services to the private sector. All the while, union power has been dramatically eroded and the public has been trained in helplessness: no matter how big the problem, we have been told, it's best to leave it to the market or billionaire philanthro-capitalists, to get out of the way, to stop trying to fix problems at their root.

That, most fundamentally, is why the historical precedents from the 1930s through to the 1950s are still useful. They remind us that another approach to profound crisis was always possible and still is today. Faced with the collective emergencies that punctuated those decades, the response was to enlist entire societies, from individual consumers to workers to large manufacturers to every level of government, in deep transitions with clear common goals.

Past problem solvers did not look for a single "silver bullet" or "killer app"; nor did they tinker and wait for the market to trickle-down fixes for them. In each instance, governments deployed a barrage of robust policy tools (from direct job creation on public infrastructure to industrial planning to public banking) all at once.

These historical chapters show us that when ambitious goals and forceful policy mechanisms are aligned, it is possible to change virtually all aspects of society on an extremely tight deadline, just as we need to do in the face of climate breakdown today. The failure to do so is a choice, not an inevitability of human nature. As Kate Marvel, a climate scientist at Columbia University and NASA's Goddard Institute for Space Studies, says, "We're not doomed (unless we choose to be)."

These precedents remind us of something equally important: we don't need to figure out every detail before we begin. Every one of these earlier mobilizations contained multiple false starts, improvisations, and course corrections. And as we will see later on, the most progressive responses happened only because of relentless pressure from organized populations. What matters is that we begin the process right away. As Greta Thunberg says, "We cannot solve an emergency without treating it like an emergency."

That does not mean we simply need a New Deal painted green, or a Marshall Plan with solar panels. We need changes of a different quality and character. We need wind and solar power that is distributed and, where possible, community owned, rather than the New Deal's highly centralized, monopolistic river-damming hydro and fossil fuel power. We need beautifully designed, racially integrated, zero-carbon urban housing, built with democratic input from communities of color—rather than the sprawling white suburbs and racially segregated urban housing projects of the postwar period. We need to devolve power and resources to Indigenous communities, smallholder farmers, ranchers, and sustainable fishing folk so they can lead a process of planting billions of trees, rehabilitating wetlands, and renewing soil—rather than handing over control of conservation to the military and federal agencies, as was overwhelmingly the case in the New Deal's Civilian Conservation Corps.

And even as we insist on naming an emergency as an emergency, we need to constantly guard against this state of emergency becoming a state of exception, in which powerful interests exploit public fear and panic to roll back hard-won rights and steamroll profitable false solutions.

In other words, we need something we've never tried, and to pull it off, we will have to recapture the sense of possibility and the can-do spirit that have been sorely missing since Ronald Reagan announced that "the nine most dangerous words in the English language are 'I'm from the government and I'm here to help.' " By reviving the historical memory of these (and other) periods of rapid collective change, we can extract both soaring inspiration and sobering warnings.

One warning from the 1930s and '40s we would be wise to remember is that when systemic crises cause political and ideological vacuums to open up, as they have today, it is not only humane and hopeful ideas like the Green New Deal that find oxygen. Violent and hateful ideas do, too. It was a truth that pierced through the first global school strikes with terrifying force on March 15, 2019.

THE SPECTER OF ECO-FASCISM

In Christchurch, New Zealand, the School Strike for Climate started in much the same way as in so many other cities and towns: rowdy students poured out of their schools in the middle of the day, holding up signs demanding a new era of climate action. Some were sweet and earnest (I STAND 4 WHAT I STAND ON), some less so (KEEP EARTH CLEAN. IT'S NOT URANUS!).

By 1 PM, about two thousand kids had made their way into Cathedral Square, at the city center, where they gathered around a makeshift stage and donated sound system to listen to speeches and music.

There were students of all ages there, and an entire Maori school had walked out together. "I was so proud of the whole of Christchurch," one of the organizers, seventeen-year-old Mia Sutherland, told me. "All of these people had been so brave. It isn't easy to walk out." The high point, she said was when the entire crowd sang a strike anthem called "Rise Up," written by twelve-year-old Lucy Gray, who had first called for the Christchurch strike. "Everyone looked so happy," Sutherland recalled, observing that this was an all-too-rare sight in a country with the highest youth suicide rates in the industrialized world.

Sutherland, an outdoorsy teen, started worrying about climate disruption when she realized it would impact the parts of the natural world that she cherished. But as she learned about sea level rise and cyclone force, and how entire Pacific nations were at risk, it became a human rights issue. "Here in New Zealand we are part of the Pacific Island family," she said. "These are our neighbors."

It wasn't only schoolkids who showed up in the square that day; so did a handful of politicians, including the mayor. But Sutherland and the other organizers decided not to let them give speeches: this was a day for kids to have the mic and politicians to listen. As emcee, it was Sutherland's job to call her fellow students up to the stage, which she did, again and again.

Just as Sutherland was psyching herself up to deliver the final testimony of the day, one of her friends gave her a tug and told her, "You have to shut it down. Now!" Sutherland was confused—had they been too loud? Surely that was their right! Just then, a police officer walked onto the stage and took the mic away from her. Everyone needs to leave the square, the officer said over the sound system. Go home. Go back to school. But stay away from Hagley Park.

A couple hundred students decided to march together to City Hall to keep the protest going. Sutherland, still confused, went to catch a bus—and that's when she saw a headline on her phone about a shooting ten minutes away from where she was standing

It would be several hours before the young strikers grasped the full horror of what had transpired that day—and why they had been told to stay away from a park near the Al Noor Mosque. We now know that at the very same time as the students' climate strike, a twenty-eight-year-old Australian man living in New Zealand drove to that mosque, walked inside, and, during Friday prayers, opened fire. After six minutes of carnage, he calmly left Al Noor, drove to another mosque, and continued his rampage. By the end, fifty people were dead, including a three-year-old child. Another would die in the hospital weeks later. An additional forty-nine were seriously injured. It was the largest massacre in modern New Zealand history.

In his manifesto (posted to multiple social media sites) and in inscriptions on his weapon, the killer expressed admiration for the men responsible for other, similar massacres: in downtown Oslo and at a Norwegian summer camp in 2011 (seventy-seven people killed); at the Emanuel African Methodist Episcopal Church in Charleston, South Carolina, in 2015 (nine people murdered); at a Quebec City mosque in 2017 (six people dead); and at Pittsburgh's Tree of Life synagogue in 2018 (eleven people murdered). Like all these other terrorists, the Christchurch shooter was obsessed with the concept of "white genocide," a supposed threat posed by the growing presence of non-white populations in majority-white nations, which he blamed on immigrant "invaders."

The horror in Christchurch was part of this clear and escalating pattern of far-right hate crimes, but it was also distinct in a couple of ways. One was the extent to which the killer planned and executed his massacre as a made-for-the-internet spectacle. Before beginning his rampage, he announced on the message board 8chan that "it's time to stop shitposting and time to make a real life effort post," as if a mass killing were merely a particularly shocking meme waiting to be shared. Then, with the help of a head-mounted camera, he proceeded to broadcast his killings live on Facebook, narrating his exploits to his imagined fans there and on YouTube and Twitter ("Alright, let's get this party started") and peppering his assault with glib references to internet in-jokes ("Remember, lads, subscribe to PewDiePie," he said, a strategic baiting of one of YouTube's top celebrities).

As his video streamed live, viewers did not report the crime in progress but, rather, cheered him on with fountains of emojis, Nazi-themed cartoon memes, and encouraging comments like "Good shooting tex." It was as if they were watching a first-person shooter game—an analogy the killer had preemptively mocked in his manifesto, where he sarcastically joked that video games had made him do it. The meta humor continued after his arrest, with the killer using his first court appearance to make the "OK" hand sign for the cameras, a move designed to set off a wave of clueless debates about whether everyone who has ever used the signal was a closet white supremacist.

At every stage, this was murder made to go viral—which of course it did, with the shooter's supporters leaping into action to play cat and mouse with censors and moderators on Facebook, YouTube, Reddit, and other sites. YouTube later reported that the snuff film was uploaded once per second in the first twenty-four hours after the attack.

The hypermediated nature of the Christchurch massacre, with the killer's obvious bid to game-ify his "real life effort post," made for an unbearable contrast with the searing reality of his horrific crime—of bullets tearing through flesh, of families stricken by grief, and of a global Muslim community sent a terrorizing message that its members were safe nowhere, not even in the sanctity of prayer.

It also made for a wrenching contrast with the youth climate strikers who had gathered at the exact same time for such a different purpose. Where the killer gleefully toyed with the lines between fact, fiction, and conspiracy, as if the very idea of truth were #FakeNews, the strikers painstakingly insisted that hard realities like accumulated greenhouse gases and carbon footprints and spiraling extinctions really did matter, and demanded that politicians close the yawning gap between their words and their actions.

Greta Thunberg had helped huge numbers of her fellow students to wake up to the seriousness of our moment in history, to stop distancing themselves from their deepest fears, and to rise up peacefully for the rights of all children. The Christchurch killer deployed extreme violence to strip whole categories of people of their humanity, even as he seemed to shrug that none of it mattered anyway.

When I spoke to her six weeks after that terrible day, Mia Sutherland was still having trouble prying the strike and the massacre apart; they had somehow fused together in her memory. "In no one's mind are they separate," she told me, her voice just above a whisper.

When intense events happen in close proximity to one another, the human mind often tries to draw connections that are not there, a phenomenon known as apophenia. But in this case, there were connections. In fact, the strike and the massacre can be understood as mirror opposite reactions to some of the same historical forces. And this relates to the other way that the Christchurch killer is distinct from the white supremacist mass murderers from whom he openly drew inspiration. Unlike them, he identifies explicitly as an "ethno-nationalist eco-fascist." In his rambling manifesto, he framed his actions as a twisted kind of environmentalism, railing against population growth and asserting that "Continued immigration into Europe is environmental warfare."

To be clear, the killer was not driven by environmental concern—his motivation was unadulterated racist hate—but ecological breakdown was one of the forces that seemed to be stoking that hatred, much as we are seeing it act as an accelerant for hatred and violence in armed conflicts around the world. My fear is that, unless something significant changes in how our societies rise to the ecological crisis, we are going to see this kind of white power eco-fascism emerge with much greater frequency, as a ferocious rationalization for refusing to live up to our collective climate responsibilities.

Much of this is due to the hard calculus warming. This is a crisis overwhelmingly created by the wealthiest strata of society: almost 50 percent of global emissions are produced by the richest 10 percent of the world's population; the wealthiest 20 percent are responsible for 70 percent. But the impacts of those emissions are hurting the poorest first and worst, forcing growing numbers of people to move, with many more on the way. A 2018 World Bank study estimates that by 2050, more than 140 million people in sub-Saharan Africa, South Asia, and Latin America will be displaced because of climate stresses, an estimate many consider conservative. Most will stay in their own countries, crowding into already overstressed cities and slums; many will try for a better life elsewhere.

In any moral universe, guided by basic human rights principles, these victims of a crisis of other people's making would be owed justice. That justice would and should take many forms. First and foremost, justice requires that the wealthiest 10–20 percent stop the underlying cause of this deepening crisis by lowering emissions as rapidly as technology allows (the premise of the Green New Deal). Justice also demands that we heed the call for a "Marshall Plan for the Earth" that Bolivia's climate negotiator called for a decade ago: to roll out resources in the Global South so communities can fortify themselves against extreme weather, pull themselves out of poverty with clean tech, and protect their ways of life wherever possible.

When protection is not possible—when the land is simply too parched to grow crops and when the seas are rising too fast to hold them back—then justice demands that we clearly recognize that all people have the human right to move and seek safety. That means they are owed asylum and status on arrival. In truth, amid so much loss and suffering, they are owed much more than that: they are owed kindness, compensation, and a heartfelt apology.

In other words, climate disruption demands a reckoning on the terrain most repellent to conservative minds: wealth redistribution, resource sharing, and reparations. And a growing number of people on the hard right realize this all too well, which is why they are developing various twisted rationales for why none of this can take place.

The first phase is to scream "socialist conspiracy" and flat out deny reality. We've been in that phase for quite some time now. That was the tack taken by Anders Breivik, the sociopath who opened fire at the Norwegian summer camp in 2011. Breivik was convinced that in addition to immigration, one of the ways that white Western culture was being weakened was through calls for Europe and the Anglosphere to pay its "climate debt." In a section of his manifesto titled "Green Is the New Red—Stop Enviro-Communism!," in which he cites several prominent climate change deniers, he casts demands for climate financing as an attempt to "'punish' European countries (US included) for capitalism and success." Climate action, he asserts, "is the new Redistribution of Wealth."

But if straight-up denial seemed a viable strategy then, nine years later (with six of those years among the ten hottest ever recorded) it is less so today. That does not mean, however, that onetime deniers are suddenly going to embrace a response to the climate crisis based on agreed-upon international frameworks. It is far more likely that many who currently claim to deny climate change will simply switch abruptly to the sinister worldview espoused by the Christchurch killer, a recognition that we are indeed facing a convulsive future and that is all the more reason for wealthy, majority-white countries to fortress their borders, as well as their identities as white Christians, and wage war on any and all "invaders."

The climate science will no longer be denied; what *will* be denied is the idea that the nations that are the largest historical emitters of carbon owe anything to the black and brown people impacted by that pollution. This will be denied based on the only rationale possible: that those non-white and non-Christian people are lesser than, are the other, are dangerous invaders.

In much of Europe and the Anglosphere, this hardening is already well under way. The European Union, Australia, and the United States have all embraced immigration policies that are variations on "prevention through deterrence." The brutal logic is to treat migrants with so much callousness and cruelty that desperate people will be deterred from seeking safety by crossing borders.

With this in mind, migrants are left to drown in the Mediterranean, or to die of dehydration in the rugged Arizona desert. And if they survive, they are put in conditions tantamount to torture:

in the Libyan camps where European countries now send the migrants who try to reach their shores; in Australia's offshore island detention camps; in a cavernous Walmart turned child jail in Texas. In Italy, if migrants do make it to a port, they are now regularly prevented from disembarking, held captive in rescue boats under conditions a court has ruled to be tantamount to kidnapping.

Canada's prime minister, meanwhile, tweets pictures of himself welcoming refugees and visiting mosques—even as his government makes massive new investments in militarizing the border and tightening the noose of the Safe Third Country Agreement, which bars asylum seekers from requesting protection at official Canadian border crossings if they are coming from the supposedly "safe" country of Trump's United States.

The goal of this fortification around Europe and the Anglosphere is all too clear: convince people to stay where they are, no matter how miserable it is, no matter how deadly. In this worldview, the emergency is not people's suffering; it is their inconvenient desire to escape that suffering.

That is why, just hours after the Christchurch massacre, Donald Trump could shrug off the surge of far-right violence and immediately change the subject to the "invasion" of migrants at the United States' southern border and his recent declaration of a "national emergency," a move meant to free up billions for a border wall. Three weeks later, Trump tweeted, "Our Country is FULL!" This followed Italy's interior minister, Matteo Salvini, responding to the arrival of a small group of migrants rescued at sea by tweeting, "Our ports were and remain CLOSED."

Murtaza Hussain, an investigative reporter who studied the Christchurch killer's manifesto closely, stresses that it is filled with ideas that are anything but marginal. His words, Hussain writes, are "both lucid and chillingly familiar. His references to immigrants as invaders find echoes in the language used by the president of the United States and far-right leaders across Europe. . . . For those wondering where [he] was radicalized, the answer is right out in the open. It is in our media and politics, where minorities, Muslims or otherwise, are vilified as a matter of course."

CLIMATE BARBARISM

The drivers of mass migration are complex: war, gang violence, sexual violence, deepening poverty. What is clear is that climate disruption is intensifying all these other crises, and it's only going to get worse as it gets hotter. But rather than helping, the wealthiest countries on the planet seem determined to deepen the crisis on every front.

They are failing to provide meaningful new aid so poorer nations can better protect themselves from weather extremes. When impoverished and debt-ridden Mozambique was pummeled by Cyclone Idai, the International Monetary Fund offered the country $118 million, a loan (not a grant) it would somehow have to pay back; the Jubilee Debt Campaign described the move as "a shocking indictment of the international community." Worse, in March 2019, Trump announced that he intended to cut $700 million in current aid to Guatemala, Honduras, and El Salvador, some of which was earmarked for programs that help farmers cope with drought. In an equally explicit expression of its priorities, in June 2018, at the start of hurricane season, the Department of Homeland Security diverted $10 million from the Federal Emergency Management Agency, which is tasked with responding to natural disasters at home, and moved it over to Immigration and Customs Enforcement, to pay for migrant detention.

Let there be no mistake: this is the dawn of climate barbarism. And unless there is a radical change not only in politics but in the underlying values that govern our politics, this is how the wealthy world is going to "adapt" to more climate disruption: by fully unleashing the toxic ideologies that rank the relative value of human lives in order to justify the monstrous discarding of huge swaths of humanity. And what starts as brutality at the border will most certainly infect societies as a whole.

These supremacist ideas are not new; nor have they ever gone away. For those of us in the Anglosphere, they are deeply embedded in the legal basis for our nations' very existence (from the Doctrine of Christian Discovery to *terra nullius*). Their power has ebbed and flowed throughout our histories, depending on what immoral behaviors demanded ideological justification. And just as these toxic ideas surged when they were required to rationalize slavery, land theft, and segregation, they are surging once more now that they are needed to justify climate recalcitrance and the barbarism at our borders.

The rapidly escalating cruelty of our present moment cannot be overstated; nor can the long-term damage to the collective psyche should this go unchallenged. Beneath the theater of some governments denying climate change and others claiming to be doing something about it while they fortress their borders from its effects, there is one overarching question facing us. In the rough and rocky future that has already begun, what kind of people are we going to be? Will we share what's left and try to look after one another? Or are we instead going to attempt to hoard what's left, look after "our own," and lock everyone else out?

In this time of rising seas and rising fascism, these are the stark choices before us. There are options besides full-blown climate barbarism, but given how far down that road we are, there is no point pretending that they are easy. It's going to take a lot more than a carbon tax or cap-and-trade. It's going to take an all-out war on pollution and poverty and racism and colonialism and despair all at the same time.

Perhaps most importantly, if we are going to avoid a future marked by more brutal scapegoating of the most vulnerable and blameless, we will need to find the fortitude to go head-to-head with the powerful players that bear the greatest responsibility for the climate crisis. Taking on the fossil fuel sector can seem impossibly daunting: it has unlimited wealth to spend lobbying politicians to pass draconian laws that target activists and to purchase public relations advertisements that pollute the public airwaves. And yet this sector is far more vulnerable to various forms of pressure than it first appears.

For the past five years, a central strategy of the climate justice movement has been to demonstrate that these companies are immoral actors whose profits are illegitimate because their core business model depends on destabilizing human civilization. That strategy has led hundreds of institutions to pledge to divest from fossil fuel stocks. More recently, the Sunrise Movement and others have focused on getting elected politicians to take a "no fossil fuel money" pledge, which well over half the contenders for the Democratic Party leadership quickly agreed to sign. If it became the policy of a governing party to refuse fossil fuel donations, and to shun fossil fuel lobbyists, the industry's hold over policymaking would be dramatically weakened. And if, under public and regulatory pressure, media outlets stopped running advertisements from fossil fuel companies, much as they stopped running tobacco ads in the past, the industry's outsized influence would be further eroded.

With less misinformation skewing debates, and a clear separation between oil and state, the path would be far clearer for the kinds of robust regulations that would quickly reign in this rogue sector, because all extractive companies function within a nonnegotiable "grow-or-die" framework: they need to constantly reassure investors that their product will be in high demand not just today but well into the future. This is why a central piece of every fossil fuel company's valuation rests not only in the projects it currently has in production but also in the oil and gas it has "in reserve"—the deposits it has discovered and purchased for development decades down the road.

According to Stephen Kretzmann, executive director of Washington-based Oil Change International, as soon as governments begin denying new exploration and drilling permits on the grounds that we need to transition rapidly to 100 percent renewable energy, investors will start to jump ship. "This defining of financial and political limits on the industry reveals their most persistent myth: that we will always need them. In fact, the reverse is true. Real climate leaders of the next decade will need to have the courage to literally withdraw all industry licenses (social, political, legal), to stop expansion of the industry urgently, and to manage a decline of production over the next few decades in a way that is fair and just towards workers and frontline communities." It may also be necessary to take over some of these companies to make sure that the remaining profits go to land and water remediation and worker pensions, rather than into investor pockets. Which in turn demands a decisive turn away from the free-market fundamentalism that has defined so much of the last half-century.

The message coming from the school strikes is that a great many young people are ready for this kind of deep change. They know all too well that the sixth mass extinction is not the only crisis they have inherited. They are also growing up in the rubble of market euphoria, in which the dreams of endlessly rising living standards have given way to rampant austerity and economic insecurity. And techno-utopianism, which imagined a frictionless future of limitless connection and community, has morphed into addiction to the algorithms of envy, relentless corporate surveillance, and spiraling online misogyny and white supremacy.

"Once you have done your homework," Greta Thunberg says, "you realize that we need new politics. We need a new economics, where everything is based on our rapidly declining and extremely limited carbon budget. But that is not enough. We need a whole new way of thinking . . . We must stop competing with each other. We need to start cooperating and sharing the remaining resources of this planet in a fair way."

Because our house is on fire, and this should come as no surprise. Built on false promises, discounted futures, and sacrificial people, it was rigged to blow from the start. It's too late to save all our stuff, but we can still save each other and a great many other species, too. Let's put out the flames and build something different in its place. Something a little less ornate, but with room for all those who need shelter and care.

Let's forge a Global Green New Deal—for everyone this time.

A HOLE IN THE WORLD

The hole at the bottom of the ocean is more than an engineering accident or a broken machine. It is a violent wound in the living organism that is Earth itself.

On April 20, 2010, BP's Deepwater Horizon offshore rig exploded in the Gulf of Mexico while it was drilling at the greatest depths ever attempted. Eleven crew members died in the fiery explosion and the wellhead ruptured, causing oil to gush uncontrollably from the ocean floor. After many failed attempts, the well was finally capped on July 15, leaving behind 4 million barrels (168 million gallons) of oil, the largest spill ever recorded in US waters.

JUNE 2010

EVERYONE GATHERED FOR THE TOWN HALL MEETING HAD BEEN REPEATEDLY instructed to show civility to the gentlemen from BP and the federal government. These fine folks had made time in their busy schedules to come to a high school gymnasium on a Tuesday night in Plaquemines Parish, Louisiana, one of many coastal communities where brown poison was slithering through the marshes, part of what has come to be described as the largest environmental disaster in US history.

"Speak to others the way you would want to be spoken to," the chair of the meeting pleaded one last time before opening the floor for questions.

And for a while the crowd, made up mostly of fishing families, showed remarkable restraint. They listened patiently to Larry Thomas, a genial BP public relations flack, as he told them that he was committed to "doing better" to process their claims for lost revenue—then passed all the details off to a markedly less friendly subcontractor. They heard out the representative from the Environmental Protection Agency as he informed them that, contrary to what they had read about the lack of testing and the product being banned in Britain, the chemical dispersant being sprayed on the oil in massive quantities was really perfectly safe.

But patience started running out by the third time Ed Stanton, a coast guard captain, took to the podium to reassure them that "the coast guard intends to make sure that BP cleans it up."

"Put it in writing!" someone shouted out. By now the air-conditioning had shut itself off and the coolers of Budweiser were running low. A shrimper named Matt O'Brien approached the mic. "We don't need to hear this anymore," he declared, hands on hips. It didn't matter what assurances they were offered because, he explained, "we just don't trust you guys!" And with that, such a loud cheer rose up from the floor you'd have thought the Oilers (the unfortunately named school football team) had scored a touchdown.

The showdown was cathartic, if nothing else. For weeks, residents had been subjected to a barrage of pep talks and extravagant promises coming from Washington, Houston, and London. Every time they turned on their TVs, there was the BP boss, Tony Hayward, offering his solemn word that he would "make it right." Or else it was President Barack Obama expressing his absolute confidence that his administration would "leave the Gulf coast in better shape than it was before," that he was "making sure" it "comes back even stronger than it was before this crisis."

It all sounded great. But for people whose livelihoods put them in intimate contact with the delicate chemistry of the wetlands, it also sounded completely ridiculous, painfully so. Once the oil coats the base of the marsh grass, as it had already done just a few miles from here, no miracle machine or chemical concoction could safely get it out. You can skim oil off the surface of open water, and you can rake it off a sandy beach, but an oiled marsh just sits there, slowly dying. The larvae of countless species for which the marsh is a spawning ground (shrimp, crab, oysters, and finfish) will be poisoned.

It was already happening. Earlier that day, I traveled through nearby marshes in a shallow-water boat. Fish were jumping in waters encircled by white boom, the strips of thick cotton and mesh that BP was using to soak up the oil. The circle of fouled material seemed to be tightening around the fish like a noose. Nearby, a red-winged blackbird perched atop a seven-foot blade of oil-contaminated marsh grass. Death was creeping up the cane; the small bird may as well have been standing on a lit stick of dynamite.

And then there is the grass itself, or the Roseau cane, as the tall, sharp blades are called. If oil seeps deeply enough into the marsh, it will kill not only the grass aboveground but also the roots. Those roots are what hold the marsh together, keeping bright green land from collapsing into the Mississippi River Delta and the Gulf of Mexico. So, places like Plaquemines Parish stand to lose not only their fisheries, but also much of the physical barrier that lessens the intensity of fierce storms like Hurricane Katrina—which could mean losing everything.

How long will it take for an ecosystem this ravaged to be "restored and made whole," as President Obama's interior secretary pledged to do? It's not at all clear that such a thing is remotely possible, at least not in a time frame we can easily wrap our heads around. The Alaskan fisheries have yet to fully recover from the 1989 *Exxon Valdez* spill, and some species of fish never returned. Government scientists now estimate that as much as a *Valdez* worth of oil may be entering the Gulf coastal waters every four days. An even worse prognosis emerges from the 1991 Gulf War spill, when an estimated eleven million barrels of oil were dumped into the Persian Gulf, the largest spill ever. That oil entered the marshland and stayed there, burrowing deeper and deeper thanks to holes dug by crabs. It's not a perfect comparison, given that so little cleanup was done, but according to a study conducted twelve years after the disaster, nearly 90 percent of the impacted muddy salt marshes and mangroves were still profoundly damaged.

We do know this. Far from being "made whole," the Gulf Coast, more than likely, will be diminished. Its rich waters and crowded skies will be less alive than they are today. The physical space many communities occupy on the map will also shrink, thanks to erosion. And the coast's legendary culture will further contract and wither. The fishing families up and down the coast don't just gather food, after all. They hold up an intricate network that includes family tradition, cuisine, music, art, and endangered languages—much like the roots of grass holding up the land in the marsh. Without fishing, these unique cultures lose their root system, the very ground on which they stand. (BP, for its part, is well aware of the limits of recovery. The company's Gulf of Mexico Regional Oil Spill Response Plan specifically instructs officials not to make "promises that property, ecology, or anything else will be restored to normal"—which is no doubt why its officials consistently favor folksy terms like "make it right.")

If Katrina pulled back the curtain on the reality of racism in

America, the BP disaster pulls back the curtain on something far more hidden: how little control even the most ingenious among us have over the awesome, intricately interconnected natural forces with which we so casually meddle. BP has spent weeks failing to plug the hole in the earth that it made. Our political leaders cannot order fish species to survive, or bottlenose dolphins not to die in droves. No amount of compensation money can replace a culture that has lost its roots. And while our politicians and corporate leaders have yet to come to terms with these humbling truths, the people whose air, water, and livelihoods have been contaminated are losing their illusions fast.

"Everything is dying," a woman said as the town hall meeting was finally coming to a close. "How can you honestly tell us that our Gulf is resilient and will bounce back? Because not one of you up here has a hint as to what is going to happen to our Gulf. You sit up here with a straight face and act like you know when you don't know."

This Gulf Coast crisis is about many things: corruption, deregulation, the addiction to fossil fuels. But underneath it all, it's about this: our culture's excruciatingly dangerous claim to have such complete understanding and command over nature that we can radically manipulate and reengineer it with minimal risk to the natural systems that sustain us. But as the BP disaster has revealed, nature is always more unpredictable than the most sophisticated mathematical and geological models imagine. During congressional testimony, BP's Hayward said, "The best minds and the deepest expertise are being brought to bear" on the crisis, and that "with the possible exception of the space program in the 1960s, it is difficult to imagine the gathering of a larger, more technically proficient team in one place in peacetime." And yet, in the face of what geologist Jill Schneiderman has described as "Pandora's well," they are like the men facing the angry crowd at the town hall meeting: they act like they know, but they don't know.

BP'S MISSION STATEMENT

In the arc of human history, the notion that nature is a machine for us to reengineer at will is a relatively recent conceit. In her groundbreaking 1980 book *The Death of Nature*, the environmental historian Carolyn Merchant reminded readers that up until the 1600s, the earth was seen as alive, usually taking the form of a mother. Europeans, like indigenous people the world over, believed the planet to be a living organism, full of life-giving powers but also wrathful tempers. There were, for this reason, strong taboos against actions that would deform or desecrate "the mother," including mining.

The metaphor changed with the unlocking of some (but by no means all) of nature's mysteries during the scientific revolution of the 1600s. With nature now cast as a machine, devoid of mystery or divinity, its component parts could be dammed, extracted, and remade with impunity. Nature still sometimes appeared as a woman, but one easily dominated and subdued. Sir Francis Bacon best encapsulated the new ethos when he wrote in the 1623 *De dignitate et augmentis scientiarum* that nature is to be "put in constraint, moulded, and made as it were new by art and the hand of man."

Those words may as well have been BP's corporate mission statement. Boldly inhabiting what the company called "the energy frontier," it dabbled in synthesizing methane-producing microbes and announced that "a new area of investigation" would be geoengineering. And of course, it bragged that at its Tiber prospect in the Gulf of Mexico, it now had "the deepest well ever drilled by the oil and gas industry," as deep under the ocean floor as jets fly overhead.

Imagining and preparing for what would happen if these experiments in altering the building blocks of life and geology went wrong occupied precious little space in the corporate imagination. As we have all discovered, after the Deepwater Horizon rig exploded, the company had no systems in place to effectively respond to this scenario. Explaining why it did not even have the ultimately unsuccessful containment dome waiting to be activated on shore, a BP spokesman, Steve Rinehart, said, "I don't think anybody foresaw the circumstance that we're faced with now." Apparently, it "seemed inconceivable" that the blowout preventer would ever fail, so why prepare?

This refusal to contemplate failure clearly came straight from the top. A year ago, CEO Hayward told a group of graduate students at Stanford University that he has a plaque on his desk that reads, IF YOU KNEW YOU COULD NOT FAIL, WHAT WOULD YOU TRY? Far from being a benign inspirational slogan, this was actually an accurate description of how BP and its competitors behaved in the real world. In recent hearings on Capitol Hill, congressman Ed Markey of Massachusetts grilled representatives from the top oil and gas companies on the revealing ways in which they had allocated resources. Over three years, they had spent "$39bn to explore for new oil and gas. Yet, the average investment in research and development for safety, accident prevention and spill response was a paltry $20m a year."

These priorities go a long way toward explaining why the initial exploration plan that BP submitted to the federal government for the ill-fated Deepwater Horizon well reads like a Greek tragedy about human hubris. The phrase "little risk" appears five times. Even if there is a spill, BP confidently predicts that thanks to "proven equipment and technology," adverse effects will be minimal. Presenting nature as a predictable and agreeable junior partner (or perhaps subcontractor), the report cheerfully explains that should a spill occur, "Currents and microbial degradation would remove the oil from the water column or dilute the constituents to background levels." The effects on fish, meanwhile, "would likely be sublethal" because of "the capability of adult fish and shellfish to avoid a spill [and] to metabolise hydrocarbons." (In BP's telling, rather than a dire threat, a spill emerges as an all-you-can-eat buffet for aquatic life.)

Best of all, should a major spill occur, there was apparently "little risk of contact or impact to the coastline" because of the company's projected speedy response (!) and "due to the distance [of the rig] to shore" (about 48 miles, or 77 kilometers). This is the most astonishing claim of all. In a gulf that often sees winds of more than 70 kilometers an hour, not to mention hurricanes, BP had so little respect for the ocean's capacity to ebb and flow, surge and heave, that it did not think oil could make a paltry 77-kilometer trip. (A shard of the exploded Deepwater Horizon showed up on a beach in Florida 306 kilometers away.)

None of this sloppiness would have been possible, however, had BP not been making its predictions to a political class eager to believe that nature had indeed been mastered. Some, like Republican Lisa Murkowski, were more eager than others. The Alaskan senator was so awestruck by the industry's four-dimensional seismic imaging that she proclaimed deep-sea drilling to have reached the very height of controlled artificiality. "It's better than Disneyland in terms of how you can take technologies and go after a resource that is thousands of years old and do so in an environmentally sound way," she told the Senate energy committee.

Drilling without thinking has of course been Republican Party policy since May 2008. When gas prices soared to unprecedented heights, the conservative leader Newt Gingrich unveiled the slogan "Drill Here, Drill Now, Pay Less," with an emphasis on the *now*. The wildly popular campaign was a cry against caution, against study, against measured action. In Gingrich's telling, drilling at home wherever the oil and gas might be—locked in Rocky Mountain shale, in the Arctic National Wildlife Refuge, and deep offshore—was a surefire way to lower the price at the pump, create jobs, and kick Arab ass all at once. In the face of this triple win, caring about the environment was for sissies: as Senator Mitch McConnell put it, "in Alabama and Mississippi and Louisiana and Texas, they think oil rigs are pretty." By the time the infamous "Drill, baby, drill" Republican National Convention rolled around in 2008, the party base was in such a frenzy for US-extracted fossil fuels that they would have bored under the convention floor if someone had brought a big enough drill.

Obama, eventually, gave in. With cosmic bad timing, just three weeks before the Deepwater Horizon blew up, the president announced that he would open up previously protected parts of the country to offshore drilling. The practice was not as risky as he had thought, he explained. "Oil rigs today generally don't cause spills. They are technologically very advanced." That wasn't enough for Sarah Palin, however, who sneered at the Obama administration's plans to conduct more studies before drilling in some areas. "My goodness, folks, these areas have been studied to death," she told the Southern Republican Leadership Conference in New Orleans just eleven days before the blowout. "Let's drill, baby, drill, not stall, baby, stall!" And there was much rejoicing.

In his congressional testimony, BP's Hayward said, "We and the entire industry will learn from this terrible event." And one might well imagine that a catastrophe of this magnitude would indeed instill BP executives and the "Drill Now" crowd with a new sense of humility. There are, however, no signs that this is the case.

The response to the disaster, at the corporate and governmental levels, has been rife with the precise brand of arrogance and overly sunny predictions that created the blowout in the first place.

"The Gulf of Mexico is a very big ocean," we heard from Hayward. "The amount of volume of oil and dispersant we are putting into it is tiny in relation to the total water volume." In other words: don't worry, she can take it. Spokesman John Curry, meanwhile, insisted that hungry microbes would consume whatever oil was in the water system because "nature has a way of helping the situation." But nature has not been playing along. The deep-sea gusher has busted through all BP's attempts at control, the so-called "top hats," "containment domes," and "junk shots." [Three months after the blowout, the wellhead was finally capped.] The ocean's winds and currents have similarly made a mockery of the lightweight booms BP has laid out to absorb the oil. "We told them," said Byron Encalade, the president of the Louisiana Oystermen Association. "The oil's gonna go over the booms or underneath the bottom." Indeed, it did. Marine biologist Rick Steiner, who has been following the cleanup closely, estimates that "70 percent or 80 percent of the booms are doing absolutely nothing at all."

And then there are the controversial chemical dispersants: more than 1.3 million gallons dumped with the company's trademark "what could go wrong?" attitude. As the angry residents at the Plaquemines Parish town hall rightly pointed out, few tests had been conducted, and there is scant research about what this unprecedented amount of dispersed oil will do to marine life. Nor is there a way to clean up the toxic mixture of oil and chemicals below the surface. Yes, fast-multiplying microbes do devour underwater oil, but in the process they also absorb the water's oxygen, creating a whole new threat to marine health.

BP had even dared to imagine that it could prevent unflattering images of oil-covered beaches and birds from escaping the disaster zone. When I was on the water with a TV crew, for instance, we were approached by another boat, whose captain asked, "Y'all work for BP?" When we said no, the response, in the open ocean, was "You can't be here, then." But, of course, these heavy-handed tactics, like all the others, have failed. There is simply too much oil in too many places. "You cannot tell God's air where to flow and go, and you can't tell water where to flow and go," I was told by environmental justice activist Debra Ramirez. It was a lesson she had learned from living in Mossville, Louisiana, surrounded by fourteen emission-spewing petrochemical plants, and watching illness spread from neighbor to neighbor.

The flow of denial shows no sign of abating. Louisiana politicians indignantly opposed Obama's temporary freeze on deepwater drilling, accusing him of killing the one big industry left standing now that fishing and tourism were in crisis. Palin mused on Facebook that "no human endeavor is ever without risk," while Texas Republican congressman John Culberson described the disaster as a "statistical anomaly." By far the most sociopathic reaction, however, came from veteran Washington commentator Llewellyn King: rather than turning away from big engineering risks, he said, we should pause in "wonder that we can build machines so remarkable that they can lift the lid off the underworld."

MAKE THE BLEEDING STOP

Thankfully, many are taking a very different lesson from the disaster, standing in wonder not at humanity's power to reshape nature, but at our powerlessness to cope with the fierce natural forces we unleash. There is something else, too. It is the feeling that the hole at the bottom of the ocean is more than an engineering accident or a broken machine. It is a violent wound in the living organism that is Earth itself. And thanks to BP's live underwater camera feed, we can all watch our planet's guts gush forth, in real time, twenty-four hours a day.

John Wathen, a conservationist with the Waterkeeper Alliance, was one of the few independent observers to fly over the spill in the early days of the disaster. After filming the thick red streaks of oil that the coast guard politely refers to as "rainbow sheen," he observed what many had felt: "The Gulf seems to be bleeding." This imagery comes up again and again in conversations and interviews. Monique Harden, an environmental rights lawyer in New Orleans, refuses to call the disaster an "oil spill" and instead says, "we are hemorrhaging." Others speak of the need to "make the bleeding stop." And I was personally struck, flying with the US Coast Guard over the stretch of ocean where the Deepwater Horizon sank, that the swirling shapes the oil made in the waves looked remarkably like cave drawings: a feathery lung gasping for air, eyes staring upward, a prehistoric bird. Messages from the deep.

And this is surely the strangest twist in the Gulf Coast saga: it seems to be waking us up to the reality that the earth never was a machine. After four hundred years of being declared dead, and in the middle of so much death, in Louisiana, the earth is coming back to life.

The experience of following the oil's progress through the ecosystem is itself a kind of crash course in deep ecology. Every day, we learn more about how what seems to be a terrible problem in one isolated part of the world actually radiates out in ways most of us could never have imagined. One day we learn that the oil could reach Cuba; then Europe. Next, we hear that fishermen all the way up the Atlantic in Prince Edward Island, Canada, are worried because the bluefin tuna they catch off their shores are born thousands of miles away in those oil-stained Gulf waters. And we learn, too, that for birds, the Gulf Coast wetlands are the equivalent of a busy airport hub. Everyone seems to have a stopover there: 110 species of migratory songbirds and 75 percent of all migratory US waterfowl.

It's one thing to be told by an incomprehensible chaos theorist that a butterfly flapping its wings in Brazil can set off a tornado in Texas. It's another to watch chaos theory unfold before your eyes. Carolyn Merchant puts the lesson like this: "The problem as BP has tragically and belatedly discovered is that nature as an active force cannot be so confined." Predictable outcomes are unusual within ecological systems, while "unpredictable, chaotic events [are] usual." And just in case we still didn't get it, a few days ago, a bolt of lightning struck a BP ship like an exclamation mark, forcing it to suspend its containment efforts. And no one dares speculate about what a hurricane would do to BP's toxic soup.

There is, it must be stressed, something uniquely twisted about this particular path to enlightenment. They say that Americans learn where foreign countries are by bombing them. Now it seems we are all learning about nature's circulatory systems by poisoning them.

In the late 1990s, an isolated indigenous group in Colombia captured world headlines with an almost *Avatar*-esque conflict. From their remote home in the Andean cloud forests, the U'wa let it be known that if Occidental Petroleum carried out plans to drill for oil on their territory, they would commit mass ritual suicide by jumping off a cliff. Their elders explained that oil is part of *ruiria*, "the blood of Mother Earth." They believe that all life, including their own, flows from *ruiria*, so pulling out the oil would bring on their destruction. (Oxy eventually withdrew from the region, saying there wasn't as much oil as it had previously thought.)

Virtually all indigenous cultures have myths about gods and spirits living in the natural world—in rocks, mountains, glaciers, forests—as did European culture before the scientific revolution. Katja Neves, an anthropologist at Concordia University, points out that the practice serves a practical purpose. Calling the earth "sacred" is another way of expressing humility in the face of forces we do not fully comprehend. When something is sacred, it demands that we proceed with caution. Even awe.

If many of us absorbed this lesson at long last, the implications would be profound. Public support for increased offshore drilling is dropping precipitously, down 22 percent from the peak of the "Drill Now" frenzy. The issue is not dead, however. Many still insist that thanks to ingenious new technology and tough new regulations, it is now perfectly safe to drill in the Arctic, where an under-ice cleanup would be infinitely more complex than the one under way in the Gulf. But perhaps this time we won't be so easily reassured, so quick to gamble with the few remaining protected havens.

Same goes for geoengineering. As climate change negotiations wear on, we should be ready to hear more from Dr. Steven Koonin, Obama's undersecretary of energy for science. He is one of the leading proponents of the idea that climate change can be combated with techno tricks like releasing sulphate and aluminum particles into the atmosphere—and of course it's all perfectly safe, just like Disneyland! He also happens to be BP's former chief scientist, the man who, just fifteen months prior to the accident, was still overseeing the technology behind BP's supposedly safe charge into deepwater drilling. Maybe this time we will opt not to let the good doctor experiment with the physics and chemistry of the earth, and choose instead to reduce our consumption and shift to renewable energies that have the virtue that, when they fail, they fail small.

The most positive possible outcome of this disaster would be not only an acceleration of renewable energy sources like wind, but a full embrace of the precautionary principle in science. The mirror opposite of Hayward's "If you knew you could not fail" credo, the precautionary principle holds that "when an activity raises threats of harm to the environment or human health," we tread carefully, as if failure were possible, even likely. Perhaps we can even get Hayward a new desk plaque to contemplate as he signs compensation checks: YOU ACT LIKE YOU KNOW, BUT YOU DON'T KNOW.

POSTSCRIPT

When I visited the Gulf Coast for this report, the spill was still ongoing and most of the lasting impacts were still unknown. Nine years later, it's clear that some of the direst predictions were proven to be correct. Research from the National Wildlife Federation indicates that three-quarters of pregnant bottlenose dolphins were not able to give birth to viable offspring in the years after the disaster. By 2015, reports indicated that the spill had been a factor in the deaths of at least five thousand mammals, many of them dolphins.

Additionally, anywhere between two to five trillion young fish were lost in the aftermath, along with more than eight billion oysters. This contributed to losses for the fishery industry of roughly $247 million in annual revenue, according to a 2015 report from the Natural Resources Defense Council (NRDC). And just as the fishing people I met worried would happen, about 12 percent of all bluefin tuna larvae in the Gulf were contaminated by oil during the 2010 spawning season, according to a study from the NRDC, with long-term population impacts that are still unknown.

The birds I saw on oiled marsh grass likely did not fare well either. Research in 2013 from Louisiana State University found that only 5 percent of sparrow nests in oiled parts of the marshland survived after the spill, compared with roughly 50 percent in marshland that was not directly impacted by oil. Findings from the Gulf of Mexico Research Initiative show that marsh grass as far as thirty feet from shore was destroyed and that large amounts of oil remained buried deep in the sediment, where it was churned up and released during Hurricane Harvey in 2012 (and will likely be released again in future disasters). According to a 2017 Florida State University study, there has been a staggering 50 percent loss of biodiversity in coastal sediment impacted by the spill.

CAPITALISM VS. THE CLIMATE

There is simply no way to square a belief system that vilifies collective action and venerates total market freedom with a problem that demands collective action on an unprecedented scale and a dramatic reining in of the market forces that created and are deepening the crisis.

NOVEMBER 2011

THERE IS A QUESTION FROM A GENTLEMAN IN THE FOURTH ROW.

He introduces himself as Richard Rothschild. He tells the crowd that he ran for county commissioner in Maryland's Carroll County because he had come to the conclusion that policies to combat global warming were actually "an attack on middle-class American capitalism." His question for the panelists, gathered in a Washington, DC, Marriott Hotel, is this: "To what extent is this entire movement simply a green Trojan horse whose belly is full with red Marxist socioeconomic doctrine?"

Here at the Heartland Institute's Sixth International Conference on Climate Change, the premier gathering for those dedicated to denying the overwhelming scientific consensus that human activity is warming the planet, this qualifies as a rhetorical question.

Like asking a meeting of German central bankers if Greeks are untrustworthy. Still, the panelists aren't going to pass up an opportunity to tell the questioner just how right he is.

Chris Horner, a senior fellow at the Competitive Enterprise Institute who specializes in harassing climate scientists with burdensome lawsuits and Freedom of Information Act fishing expeditions, angles the table mic over to his mouth. "You can believe this is about the climate," he says darkly, "and many people do, but it's not a reasonable belief." Horner, whose prematurely silver hair makes him look like a right-wing Anderson Cooper, likes to invoke Saul Alinsky: "The issue isn't the issue." The issue, apparently, is that "no free society would do to itself what this agenda requires. . . . The first step to that is to remove these nagging freedoms that keep getting in the way."

Claiming that climate change is a plot to steal American freedom is rather tame by Heartland standards. Over the course of this two-day conference, I will learn that Obama's campaign promise to support locally owned biofuels refineries was really about "green communitarianism," akin to the "Maoist" scheme to put "a pig iron furnace in everybody's backyard" (the Cato Institute's Patrick Michaels); that climate change is "a stalking horse for National Socialism" (former Republican senator and retired astronaut Harrison Schmitt); and that environmentalists are like Aztec priests, sacrificing countless people to appease the gods and change the weather (Marc Morano, editor of the denialists' go-to website, ClimateDepot).

Most of all, however, I will hear versions of the opinion expressed by the county commissioner in the fourth row: that climate change is a Trojan horse designed to abolish capitalism and replace it with some kind of eco-socialism. As conference speaker Larry Bell succinctly puts it in his new book, *Climate of Corruption*, climate change "has little to do with the state of the environment and much to do with shackling capitalism and transforming the American way of life in the interests of global wealth redistribution."

Yes, sure, there is a pretense that the delegates' rejection of climate science is rooted in serious disagreement about the data. And the organizers go to some lengths to mimic credible scientific conferences, calling the gathering "Restoring the Scientific Method" and even adopting the organizational acronym ICCC, a mere one letter off from the world's leading authority on climate change, the Intergovernmental Panel on Climate Change (IPCC). But the scientific theories presented here are old, and long discredited. And no attempt is made to explain why each speaker seems to contradict the next. (Is there no warming, or is there warming but it's not a problem? And if there is no warming, then what's all this talk about sunspots causing temperatures to rise?)

In truth, several members of the mostly elderly audience seem to doze off while the temperature graphs are projected. They come to life only when the rock stars of the movement take the stage—not the C-team scientists but the A-team ideological warriors like Morano and Horner. This is the true purpose of the gathering: providing a forum for die-hard denialists to collect the rhetorical baseball bats with which they will club environmentalists and climate scientists in the weeks and months to come. The talking points first tested here will jam the comment sections beneath every article and YouTube video that contains the phrase "climate change" or "global warming." They will also exit the mouths of hundreds of right-wing commentators and politicians—from Republican presidential candidates like Rick Perry and Michele Bachmann all the way down to county commissioners like Richard Rothschild. In an interview outside the sessions, Joseph Bast, president of the Heartland Institute, proudly takes credit for "thousands of articles and op-eds and speeches . . . that were informed by or motivated by somebody attending one of these conferences."

The Heartland Institute, a Chicago-based think tank devoted to "promoting free-market solutions," has been holding these confabs since 2008, sometimes twice a year. And the strategy appears to be working. At the end of day one, Morano—whose claim to fame is having broken the Swift Boat Veterans for Truth story that sank John Kerry's 2004 presidential campaign—leads the gathering through a series of victory laps. Cap-and-trade: dead! Obama at the Copenhagen summit: failure! The climate movement: suicidal! He even projects a couple of quotes from climate activists beating up on themselves (as progressives do so well) and exhorts the audience to "celebrate!"

There were no balloons or confetti descending from the rafters, but there may as well have been.

When public opinion on the big social and political issues changes, the trends tend to be relatively gradual. Abrupt shifts, when they come, are usually precipitated by dramatic events. Which is why pollsters are so surprised by what happened to perceptions about climate change in the United States over a span of just four years. A 2007 Harris poll found that 71 percent of Americans believed that the continued burning of fossil fuels would cause the climate to change. By 2009, the figure had dropped to 51 percent. In June 2011, the number of Americans who agreed was down to 44 percent—well under half the population. According to Scott Keeter, director of survey research at the Pew Research Center for the People and the Press, this is "among the largest shifts over a short period of time seen in recent public opinion history."*

Even more striking, this shift has occurred almost entirely at one end of the political spectrum. As recently as 2008 (the year Newt Gingrich did a climate change TV spot with Nancy Pelosi), the issue still had a veneer of bipartisan support in the United States. Those days are decidedly over. Today, 70–75 percent of self-identified Democrats and liberals believe humans are changing the climate, a level that has remained stable or risen slightly over the past decade. In sharp contrast, Republicans, particularly Tea Party members, have overwhelmingly chosen to reject the scientific consensus. In some regions, only about 20 percent of self-identified Republicans accept the science.†

Equally significant has been a shift in emotional intensity. Climate change used to be something most everyone said they cared about—just not all that much. When Americans were asked to rank

*The numbers have rebounded since and were shifting rapidly in early 2019. A January 2019 study from the Yale Program on Climate Change Communication found that 72 percent of Americans described climate change as "personally important" to them—a nine-point increase since March 2018. A clear majority also understood climate change to be caused mainly by human activity. The study also found that "Nearly half of Americans (46%) say they have personally experienced the effects of global warming, an increase of 15 percentage points since March 2015." Also significant, 2017 polling from Pew Research found that 65 percent of Americans support expanding non-fossil fuel energy sources, while only 27 percent support doubling down on fossil fuels.

†The partisan divide remains stark, with only 26 percent of conservative Republicans believing the scientific consensus on climate change. However, among self-identified liberal/moderate Republicans, there has been a significant drop in denial, with 55 percent now acknowledging humanity's role in global warming, according to a Yale study.

their political concerns in order of priority, climate change would reliably come in last.*

But now there is a significant cohort of Republicans who care passionately, even obsessively, about climate change—though what they care about is exposing it as a "hoax" being perpetrated by liberals to force them to change their lightbulbs, live in Soviet-style tenements, and surrender their SUVs. For these right-wingers, opposition to climate change has become as central to their worldview as low taxes, gun ownership, and opposition to abortion. Many climate scientists report receiving death threats, as do authors of articles on subjects as seemingly innocuous as energy conservation. (As one letter writer put it to Stan Cox, author of a book critical of air-conditioning, "You can pry my thermostat out of my cold dead hands.")

This culture war intensity is the worst news of all because when you challenge a person's position on an issue core to his or her identity, facts and arguments are seen as little more than further attacks, easily deflected. (The deniers have even found a way to dismiss a new study confirming the reality of global warming that was partially funded by the conservative billionaires the Koch brothers and led by a scientist sympathetic to the "skeptic" position.)

The effects of this emotional intensity have been on full display in the race to lead the Republican Party. Days into his presidential campaign, with his home state literally burning up with wildfires, Texas governor Rick Perry delighted the base by declaring that

*This may be the biggest recent shift of all: an early 2019 poll from the Pew Research Center indicated that 44 percent of US voters think climate change should be a top priority, up from 26 percent in 2011. Most remarkably, a CNN poll conducted in April 2019 suggests that climate change is now the top issue of importance for registered Democratic Party voters ahead of the presidential primaries, rating even higher than health care.

climate scientists were manipulating data "so that they will have dollars rolling into their projects." Meanwhile, the campaign of the only candidate to consistently defend climate science, Jon Huntsman, was dead on arrival. And part of what rescued Mitt Romney's campaign was his flight from earlier statements supporting the scientific consensus on climate change.

But the effects of the right-wing climate conspiracies reach far beyond the Republican Party. The Democrats have mostly gone mute on the subject, not wanting to alienate independents. And the media and culture industries have followed suit. In 2007, celebrities were showing up at the Academy Awards in hybrids. That same year, *Vanity Fair* launched an annual green issue, and the three major US television networks ran 147 stories on climate change. No longer. In 2010 the major networks ran just 32 climate change stories; limos are back in style at the Academy Awards; and the "annual" *Vanity Fair* green issue hasn't been seen since 2008.

This uneasy silence has persisted through the end of the hottest decade in recorded history and yet another summer of freak natural disasters and record-breaking heat worldwide. Meanwhile, the fossil fuel industry is rushing to make multibillion-dollar investments in new infrastructure to extract oil, natural gas, and coal from some of the dirtiest and highest-risk sources on the continent (the $7 billion Keystone XL pipeline being only the highest-profile example). In the Alberta tar sands, in the Beaufort Sea, in the gas fields of Pennsylvania and the coalfields of Wyoming and Montana, the industry is betting big that the prospect of serious climate legislation is as good as dead.

If the interred carbon these projects are poised to extract is released into the atmosphere, the chance of triggering catastrophic climate change will increase dramatically (mining all the oil in the

Alberta tar sands alone, says NASA's James Hansen, would be "essentially game over" for the climate).

All this means that the climate movement needs to have one hell of a comeback. For this to happen, the left is going to have to learn from the right. Denialists gained traction by making climate about economics: action will destroy capitalism, they have claimed, killing jobs and sending prices soaring. But at a time when a growing number of people agree with the protesters at Occupy Wall Street, many of whom argue that capitalism-as-usual is itself the cause of job insecurity and debt peonage, there is a wide-open opportunity to seize the economic terrain from the right. This would require making a persuasive case that the real solutions to the climate crisis are also our best hope of building a fairer and much more enlightened economic system—one that closes deep inequalities, strengthens and transforms the public sphere, generates plentiful, dignified work, and radically reins in corporate power. It would also require a shift away from the notion that climate action is just one issue on a laundry list of worthy causes vying for progressive attention. Just as climate denialism has become a core identity issue on the right, utterly entwined with defending current systems of power and wealth, the scientific reality of climate change must, for progressives, occupy a central place in a coherent narrative about the perils of unrestrained greed and the need for real alternatives.

Building such a transformative movement may not be as hard as it first appears. Indeed, if you ask the Heartlanders, climate change makes some kind of left-wing revolution virtually inevitable, which is precisely why they are so determined to deny its reality. Perhaps we should listen to their theories more closely—they might just understand something most of us still do not get.

• • •

The deniers did not decide that climate change is a left-wing conspiracy by uncovering some covert socialist plot. They arrived at this analysis by taking a hard look at what it would take to lower global emissions as drastically and as rapidly as climate science demands. They have concluded that this can be done only by radically reordering our economic and political systems in ways antithetical to their "free-market" belief system. As British blogger and Heartland regular James Delingpole has pointed out, "Modern environmentalism successfully advances many of the causes dear to the left: redistribution of wealth, higher taxes, greater government intervention, regulation." Heartland's Bast puts it even more bluntly: For the left, "Climate change is the perfect thing. . . . It's the reason why we should do everything [the left] wanted to do anyway."

Here's my inconvenient truth: they aren't wrong. Before I go any further, let me be absolutely clear: as 97 percent of the world's climate scientists attest, the Heartlanders are completely wrong about the science. The heat-trapping gases released into the atmosphere through the burning of fossil fuels and the clearing of our forests are already causing temperatures to increase. If we are not on a radically different energy path by the end of this decade, we are in for a world of pain.

But when it comes to the political consequences of those scientific findings, specifically the kind of deep changes required not just to our energy consumption but to the underlying logic of our economic system, the crowd gathered at the Marriott Hotel may be considerably less in denial than a lot of professional environmentalists, the ones who paint a picture of global warming Armageddon and then assure us that we can avert catastrophe by buying "green" products and creating clever markets in pollution.

The fact that the earth's atmosphere cannot safely absorb the amount of carbon we are pumping into it is a symptom of a much larger crisis, one born of the central fiction on which our economic model is based: that nature is limitless, that we will always be able to find more of what we need, and that if something runs out, it can be seamlessly replaced by another resource that we can endlessly extract. And it is not just the atmosphere that we have exploited beyond its capacity to recover—we are doing the same to the oceans, to freshwater, to topsoil, and to biodiversity. The expansionist, extractive mind-set that has so long governed our relationship to nature is what the climate crisis calls into question so fundamentally. The abundance of scientific research showing we have pushed nature beyond its limits demands not just green products and market-based solutions, but a new civilizational paradigm, one grounded not in dominance over nature but in respect for natural cycles of renewal—and acutely sensitive to natural limits, including the limits of human intelligence.

So, in a way, Chris Horner was right when he told his fellow Heartlanders that climate change isn't "the issue." In fact, it isn't an issue at all. Climate change is a message, one that is telling us that many of Western culture's most cherished ideas are no longer viable. These are profoundly challenging revelations for all of us raised on Enlightenment ideals of progress, unaccustomed to having our ambitions confined by natural boundaries. And this is true for the statist left as well as the neoliberal right.

While Heartlanders like to invoke the specter of communism to terrify Americans about climate action (former Czech president Vaclav Klaus, a Heartland conference favorite, says that attempts to prevent global warming are akin to "the ambitions of communist central planners to control the entire society"), the reality is that Soviet-era state socialism was a disaster for the climate. It devoured resources with as much enthusiasm as capitalism, and spewed waste just as recklessly: before the fall of the Berlin Wall, Czechs and Russians had even higher carbon footprints per capita than their counterparts in Britain, Canada, and Australia. And while some point to the dizzying expansion of China's renewable energy programs to argue that only centrally controlled regimes can get the green job done, China's command-and-control economy continues to be harnessed to wage an all-out war with nature, through massively disruptive mega-dams, superhighways, and extraction-based energy projects, particularly coal.*

It is true that responding to the climate threat requires a willingness to engage in industrial planning and strong government action at all levels. But some of the most successful climate solutions are ones that steer these interventions to systematically disperse and devolve power and control to the community level, whether through community-controlled renewable energy, ecological agriculture, or transit systems genuinely accountable to their users.

Here is where the Heartlanders have good reason to be afraid: arriving at these new systems is going to require shredding the free-market ideology that has dominated the global economy for more than three decades. What follows is a broad strokes look at

*Columbia University's Center on Global Energy Policy reports some encouraging recent trends: China is now the world leader in wind, solar, and hydro power. Coal consumption, which had been increasing steadily, dropped by 3–4 percent in 2017. However, though public anger at toxic air pollution has succeeded in shutting down many coal-fired power plants inside China and blocking the construction of many new ones, a reported one hundred new plants are being constructed in other countries with Chinese involvement. In other words, just as North America and Europe outsourced much of their emissions to China along with their manufacturing, now China is outsourcing a portion of its emissions to poorer parts of the world.

what a serious climate agenda would mean in the following six arenas: public infrastructure, economic planning, corporate regulation, international trade, consumption, and taxation. For hard-right ideologues like those gathered at the Heartland conference, the results are nothing short of intellectually cataclysmic.

1. REVIVING AND REINVENTING THE PUBLIC SPHERE

After years of recycling, carbon offsetting, and lightbulb changing, it is obvious that individual action will never be an adequate response to the climate crisis. Climate change is a collective problem, and it demands collective action. One of the key areas in which this collective action must take place is big-ticket investments designed to reduce our emissions on a mass scale. That means subways, streetcars, and light-rail systems that are not only everywhere but affordable to everyone, and maybe even free; energy-efficient affordable housing along those transit lines; smart electrical grids carrying renewable energy; and a massive research effort to ensure that we are using the best methods possible.

The private sector is ill suited to provide most of these services because they require large up-front investments, and if they are to be genuinely accessible to all, some very well may not be profitable. They are, however, decidedly in the public interest, which is why they should come from the public sector.

Traditionally, battles to protect the public sphere are cast as conflicts between irresponsible leftists who want to spend without limit and practical realists who understand that we are living beyond our economic means. But the gravity of the climate crisis cries out for a radically new conception of realism, and a very different understanding of limits. Government budget deficits are not nearly as dangerous as the deficits we have created in vital and complex natural systems. Changing our culture to respect those limits will require all of our collective muscle—to get ourselves off fossil fuels and to shore up communal infrastructure for the coming storms.

2. REMEMBERING HOW TO PLAN

In addition to reversing the thirty-year privatization trend, a serious response to the climate threat involves recovering an art that has been relentlessly vilified during these decades of market fundamentalism: planning. Lots and lots of planning. Industrial planning. Land use planning. And not just at the national and international levels. Every city and community in the world needs a plan for how it is going to transition away from fossil fuels, what the Transition Town movement calls an "energy descent action plan." In the cities and towns that have taken this responsibility seriously, the process has opened rare spaces for participatory democracy, with neighbors packing consultation meetings at city halls to share ideas about how to reorganize their communities to lower emissions and build in resilience for tough times ahead.

Climate change demands other forms of planning as well, particularly for workers whose jobs will become obsolete as we wean ourselves off fossil fuels. A few "green jobs" training sessions aren't enough. These workers need to know that real jobs will be waiting for them on the other side. That means bringing back the idea of planning our economies based on collective priorities rather than corporate profitability—giving laid-off employees of car plants and coal mines the tools and resources to get equally secure jobs making subway cars, installing wind turbines and cleaning up extraction sites, to cite just a few examples. Some of this will be in the private sector, some in the public realm, and some in cooperatives, with

Cleveland's worker-run green co-ops serving as a possible model.

Agriculture, too, will have to see a revival in planning if we are to address the triple crisis of soil erosion, extreme weather, and dependence on fossil fuel inputs. Wes Jackson, the visionary founder of the Land Institute in Salina, Kansas, has been calling for "a fifty-year farm bill." That's the length of time he and his collaborators Wendell Berry and Fred Kirschenmann estimate it will take to conduct the research and develop the infrastructure to replace many soil-depleting annual grain crops (grown in monocultures) with perennial crops (grown in polycultures). Because perennials don't need to be replanted every year, their long roots do a much better job of storing scarce water, holding soil in place and sequestering carbon. Polycultures are also less vulnerable to pests and to being wiped out by the extreme weather that is already locked in. Another bonus: this type of farming is much more labor intensive than industrial agriculture, which means that farming can once again be a substantial source of employment in long neglected rural communities.

Outside the Heartland conference and like-minded gatherings, the return of planning is nothing to fear. The thirty-odd-year experiment in deregulated, Wild West economics is failing the vast majority of people around the world. These systemic failures are precisely why so many are in open revolt against their elites, demanding living wages, an end to corruption, and real democracy. Climate change doesn't conflict with demands for a new kind of economy. Rather, it adds to them an existential imperative.

3. REINING IN CORPORATIONS

A key piece of the planning we must undertake involves the rapid reregulation of the corporate sector. Much can be done with incentives: subsidies for renewable energy and responsible land stewardship, for instance. But we are also going to have to get back into the habit of barring outright dangerous and destructive behavior. That means getting in the way of corporations on multiple fronts, from imposing strict caps on the amount of carbon corporations can emit, to banning new coal-fired power plants, to cracking down on industrial feedlots, to phasing out dirty-energy extraction projects (starting with canceling new pipelines and other infrastructure projects that, if built, would lock in expansion plans).

Only a very small sector of the population sees any restriction on corporate or consumer choice as leading down Hayek's road to serfdom—and not coincidentally, it is precisely this sector of the population that is at the forefront of climate change denial.

4. RELOCALIZING PRODUCTION

If strictly regulating corporations to respond to climate change sounds somewhat radical it's because, since the beginning of the 1980s, it has been an article of faith that the role of government is to get out of the way of the corporate sector, and nowhere more so than in the realm of international trade. The devastating impacts of free trade on manufacturing, local business, and farming are well known. But perhaps the atmosphere has taken the hardest hit of all. The cargo ships, jumbo jets, and heavy trucks that haul raw resources and finished products across the globe devour fossil fuels and spew greenhouse gases. And the cheap goods being produced—made to be replaced, almost never repaired—are consuming a huge range of other nonrenewable resources while producing far more waste than can be safely absorbed.

This model is so wasteful, in fact, that it cancels out the modest gains that have been made in reducing emissions many times over.

For instance, the *Proceedings of the National Academy of Sciences* recently published a study of the emissions from industrialized countries that signed the Kyoto Protocol. It found that while they had stabilized, that was partly because international trade had allowed these countries to move their dirty production to places like China. The researchers concluded that the rise in emissions from goods produced in developing countries but consumed in industrialized ones was *six times* greater than the emissions savings of industrialized countries.

In an economy organized to respect natural limits, the use of energy-intensive long-haul transport would need to be rationed—reserved for those cases where goods cannot be produced locally or where local production is more carbon-intensive. (For example, growing food in greenhouses in cold parts of the United States is often more energy-intensive than growing it in the South and shipping it by light rail.)

Climate change does not demand an end to trade. But it does demand an overhaul of the reckless form of "free trade" that governs every bilateral trade agreement and the World Trade Organization. If done thoughtfully and carefully, this is more good news—for unemployed workers, for farmers unable to compete with cheap imports, for communities that have seen their manufacturers move offshore and their local businesses replaced with big-box stores. But the challenge this poses to the capitalist project should not be underestimated: it represents the reversal of the thirty-year trend of removing every possible limit on corporate power.

5. ENDING THE CULT OF SHOPPING

The past three decades of free trade, deregulation and privatization were not only a result of greedy people wanting greater corporate profits. They were also a response to the "stagflation" of the 1970s, which created intense pressure to find new avenues for rapid economic growth. The threat was real: within our current economic model, a drop in production is, by definition, a crisis—a recession or, if deep enough, a depression, with all the desperation and hardship that these words imply.

This growth imperative is why conventional economists reliably approach the climate crisis by asking the question, How can we reduce emissions while maintaining robust GDP growth? The usual answer is "decoupling," the idea that renewable energy and greater efficiencies will allow us to sever economic growth from its environmental impact. And "green growth" advocates such as Thomas Friedman tell us that the process of developing new green technologies and installing green infrastructure can provide a huge economic boost, sending GDP soaring and generating the wealth needed to "make America healthier, richer, more innovative, more productive, and more secure."

But here is where things get complicated. There is a growing body of economic research on the conflict between unchecked economic growth and sound climate policy, led by ecological economist Herman Daly at the University of Maryland, Peter Victor at York University, Tim Jackson of the University of Surrey, and environmental law and policy expert Gus Speth. All raise serious questions about the feasibility of industrialized countries making the deep emissions cuts demanded by science [getting to net zero before mid-century] while continuing to grow their economies at even today's sluggish rates. As Victor and Jackson argue, greater efficiencies simply cannot keep up with the pace of growth, in part because greater efficiency is almost always accompanied by more consumption, reducing or even canceling out the gains (often called the "Jevons paradox"). And so long as the savings resulting from greater energy and material efficiencies are simply plowed back into further exponential expansion of the economy, reduction in total emissions will be thwarted. As Jackson argues in *Prosperity Without Growth*, "Those who promote decoupling as an escape route from the dilemma of growth need to take a closer look at the historical evidence—and at the basic arithmetic of growth."

The bottom line is that an ecological crisis that has its roots in the overconsumption of natural resources must be addressed not just by improving the efficiency of our economies, but also by reducing the amount of material stuff that the wealthiest 20 percent of people on the planet consume. Yet that idea is anathema to the large corporations that dominate the global economy, which are controlled by footloose investors who demand ever-greater profits year after year. We are therefore caught in the untenable bind of, as Jackson puts it, "trash the system or crash the planet."

The way out is to embrace a managed transition to another economic paradigm, using all the tools of planning just discussed. Increases in consumption should be reserved for those around the world still pulling themselves out of poverty. Meanwhile, in the industrialized world, those sectors that are not governed by the drive for increased yearly profit (the public sector, co-ops, local businesses, nonprofits) would expand their share of overall economic activity, as would those sectors with minimal ecological impacts but outsized benefits for well-being (such as teaching, the caregiving professions and leisure activities). A great many jobs could be created this way. But the role of the corporate sector, with its structural demand for increased sales and profits, would have to contract, particularly those segments whose fortunes are inextricable from resource extraction.

So, when the Heartlanders react to evidence of human-induced climate change as if capitalism itself were coming under threat, it's not because they are paranoid. It's because they are paying attention.

6. TAXING THE RICH AND FILTHY

About now a sensible reader would be asking, How on earth are we going to pay for all this? The old answer would have been easy: we'll grow our way out of it. Indeed, one of the major benefits of a growth-based economy for elites is that it allows them to constantly defer demands for economic justice, claiming that if we keep growing the pie, eventually there will be enough for everyone. That was always a lie, as the current inequality crisis reveals, but in a world hitting multiple ecological limits, it is a nonstarter. So, the only way to finance a meaningful response to the ecological crisis is to go where the money is.

That means taxing carbon, and financial speculation. It means increasing taxes on corporations and the wealthy, cutting bloated military budgets, and eliminating absurd subsidies to the fossil fuel industry ($20 billion annually in the United States alone). And governments will have to coordinate their responses so that corporations will have nowhere to hide. (This kind of robust international regulatory architecture is what Heartlanders mean when they warn that climate change will usher in a sinister "world government.")

Most of all, however, we need to go after the profits of the corporations most responsible for getting us into this mess. The top five oil companies made $900 billion in profits in the past decade; ExxonMobil alone can clear $10 billion in profits in a single quarter. For years, these companies have pledged to use their profits to invest in a shift to renewable energy (BP's "Beyond Petroleum" rebranding being the highest-profile example). But according to a study by the Center for American Progress, just 4 percent of the big five's 2008 combined $100 billion profits went to "renewable and alternative energy ventures." Instead, they continue to pour their profits into shareholder pockets, outrageous executive pay, and new technologies designed to extract even dirtier and more dangerous fossil fuels. Plenty of money has also gone to paying lobbyists to beat back every piece of climate legislation that has reared its head, and to fund the denier movement gathered at the Marriott Hotel.

Just as tobacco companies have been obliged to pay the costs of helping people to quit smoking, and BP has had to pay for a large portion of the cleanup in the Gulf of Mexico, it is high time for the "polluter pays" principle to be applied to climate change. Beyond higher taxes on polluters, governments will have to negotiate much higher royalty rates so that less fossil fuel extraction would raise more public revenue to pay for the shift to our postcarbon future (and the steep costs of climate change already upon us). Since corporations can be counted on to resist any new rules that cut into their profits, nationalization, the greatest free-market taboo of all, cannot be off the table.

When Heartlanders claim, as they so often do, that climate change is a plot to "redistribute wealth" and wage class war, these are the types of policies they most fear. They also understand that once the reality of climate change is recognized, wealth will have to be transferred not just within wealthy countries but also from the rich countries whose emissions created the crisis to poorer ones that are on the front lines of its effects. Indeed, what makes conservatives (and plenty of liberals) so eager to bury the UN climate negotiations is that they have revived an anti-colonial courage in parts of the developing world that many thought was gone for good. Armed with irrefutable scientific facts about who is responsible for global warming and who is suffering its effects first and worst, countries like Bolivia and Ecuador are attempting to shed the mantle of "debtor" thrust upon them by decades of International Monetary Fund and World Bank loans and are declaring themselves creditors—owed not just money and technology to cope with climate change but also "atmospheric space" in which to develop.

So, let's summarize. Responding to climate change requires that we break every rule in the free-market playbook and that we do so with great urgency. We will need to rebuild the public sphere, reverse privatizations, relocalize large parts of economies, scale back overconsumption, bring back long-term planning, heavily regulate and tax corporations, maybe even nationalize some of them, cut military spending, and recognize our debts to the Global South. Of course, none of this has a hope in hell of happening unless it is accompanied by a massive, broad-based effort to radically reduce the influence that corporations have over the political process. That means, at a minimum, publicly funded elections and stripping corporations of their status as "people" under the law. In short, climate change supercharges the preexisting case for virtually every progressive demand on the books, binding them into a coherent agenda based on a clear scientific imperative.

More than that, climate change implies the biggest political "I told you so" since Keynes predicted German backlash from the Treaty of Versailles. Marx wrote about capitalism's "irreparable rift" with "the natural laws of life itself," and many on the left have argued that an economic system built on unleashing the voracious appetites of capital would overwhelm the natural systems on which life depended. And, of course, indigenous peoples were issuing warnings about the dangers of disrupting natural cycles long before that. The fact that the airborne waste of industrial capitalism is causing the planet to warm, with potentially cataclysmic results, means that, well, the naysayers were right. And the people who said, "Hey, let's get rid of all the rules and watch the magic happen" were disastrously, catastrophically wrong.

There is no joy in being right about something so terrifying. But for progressives, there is responsibility in it, because it means that our ideas, informed by indigenous teachings as well as by the failures of industrial state socialism, are more important than ever. It means that a green-left worldview, which rejects mere reformism and challenges the centrality of profit in our economy, offers humanity's best hope of overcoming these overlapping crises.

But imagine, for a moment, how all this looks to a guy like Heartland president Joseph Bast, who studied economics at the University of Chicago and described his personal calling to me as "freeing people from the tyranny of other people." To him, it looks like the end of the world. It's not, of course. But it is, for all intents and purposes, the end of *his* world. Climate change detonates the ideological scaffolding on which contemporary conservatism rests. There is simply no way to square a belief system that vilifies collective action and venerates total market freedom with a problem that demands collective action on an unprecedented scale and a dramatic reining in of the market forces that created and are deepening the crisis.

At the Heartland conference, where everyone from the Ayn Rand Institute to the Heritage Foundation has a table hawking books and pamphlets, these anxieties are close to the surface. Bast is forthcoming about the fact that Heartland's campaign against climate science grew out of fear about the policies that the science would require. "When we look at this issue, we say, This is a recipe for massive increase in government. . . . Before we take this step, let's take another look at the science. So conservative and libertarian groups, I think, stopped and said, Let's not simply accept this as an article of faith; let's actually do our own research." This is a crucial point to understand: it is not opposition to the scientific facts of climate change that drives denialists, but rather, opposition to the real-world implications of those facts.

What Bast is describing, albeit inadvertently, is a phenomenon receiving a great deal of attention from a growing subset of social scientists trying to explain the dramatic shifts in belief about climate change. Researchers with Yale's Cultural Cognition Project have found that political/cultural worldview explains "individuals' beliefs about global warming more powerfully than any other individual characteristic."

Those with strong "egalitarian" and "communitarian" worldviews (marked by an inclination toward collective action and social justice, concern about inequality, and suspicion of corporate power) overwhelmingly accept the scientific consensus on climate change. On the other hand, those with strong "hierarchical" and "individualistic" worldviews (marked by opposition to government assistance for the poor and minorities, strong support for industry, and a belief that we all get what we deserve) overwhelmingly reject the scientific consensus.

For example, among the segment of the US population that displays the strongest "hierarchical" views, only 11 percent rate climate change as a "high risk," compared with 69 percent of the segment displaying the strongest "egalitarian" views. Yale law professor Dan Kahan, the lead author on this study, attributes this tight correlation between "worldview" and acceptance of climate science to "cultural cognition." This refers to the process by which all of us, regardless of political leanings, filter new information in ways designed to protect our "preferred vision of the good society." As Kahan explained in *Nature*, "People find it disconcerting to believe that behavior that they find noble is nevertheless detrimental to society, and behavior that they find base is beneficial to it. Because accepting such a claim could drive a wedge between them and their peers, they have a strong emotional predisposition to reject it." In other words, it is always easier to deny reality than to watch your worldview get shattered, a fact that was as true of die-hard Stalinists at the height of the purges as it is of libertarian climate deniers today.

When powerful ideologies are challenged by hard evidence from the real world, they rarely die off completely. Rather, they become cultlike and marginal. A few true believers always remain to tell one another that the problem wasn't with the ideology; it was the weakness of leaders who did not apply the rules with sufficient rigor. We have these types on the Stalinist left, and they exist as well on the neo-Nazi right. By this point in history, free-market fundamentalists should be exiled to a similarly marginal status, left to fondle their copies of *Free to Choose* and *Atlas Shrugged* in obscurity. They are saved from this fate only because their ideas about minimal government, no matter how demonstrably at war with reality, remain so profitable to the world's billionaires that they are kept fed and clothed in think tanks by the likes of Charles and David Koch, and ExxonMobil.

This points to the limits of theories like cultural cognition. The deniers are doing more than protecting their cultural worldview—they are protecting powerful interests that stand to gain enormously from muddying the waters of the climate debate. The ties between the deniers and those interests are well known and well documented. Heartland has received more than $1 million from

ExxonMobil together with foundations linked to the Koch brothers and Richard Mellon Scaife (possibly much more, but the think tank has stopped publishing its donors' names, claiming the information was distracting from the "merits of our positions").*

And scientists who present at Heartland climate conferences are almost all so steeped in fossil fuel dollars that you can practically smell the fumes. To cite just two examples, the Cato Institute's Patrick Michaels, who gave the conference keynote, once told CNN that 40 percent of his consulting company's income comes from oil companies, and who knows how much of the rest comes from coal. A Greenpeace investigation into another one of the conference speakers, astrophysicist Willie Soon, found that between 2002 and 2011, 100 percent of his new research grants had come from fossil fuel interests. And fossil fuel companies are not the only economic interests strongly motivated to undermine climate science. If solving this crisis requires the kinds of profound changes to the economic order that I have outlined, then every major corporation benefiting from loose regulation, free trade, and low taxes has reason to fear.

With so much at stake, it should come as little surprise that climate deniers are, on the whole, those most invested in our highly unequal and dysfunctional economic status quo. One of the most interesting findings in studies on climate perceptions is the clear connection between a refusal to accept the science of climate change, and social and economic privilege. Overwhelmingly, climate deniers

*This is a systemic problem. According to a 2014 study published in *Climate Change*, the denial-espousing think tanks and other advocacy groups making up what sociologist Robert Brulle calls the "climate change countermovement" are collectively pulling in more than $900 million per year for their work on a variety of right-wing causes, most of it in the form of "dark money," funds from conservative foundations that cannot be fully traced.

are not only conservative but also white and male, a group with higher-than-average incomes. And they are more likely than other adults to be highly confident in their views, no matter how demonstrably false. A much-discussed paper on this topic by Aaron McCright and Riley Dunlap (memorably titled "Cool Dudes") found that confident conservative white men, as a group, were almost six times as likely to believe climate change "will never happen" than the rest of the adults surveyed. McCright and Dunlap offer a simple explanation for this discrepancy: "Conservative white males have disproportionately occupied positions of power within our economic system. Given the expansive challenge that climate change poses to the industrial capitalist economic system, it should not be surprising that conservative white males' strong system-justifying attitudes would be triggered to deny climate change."

But deniers' relative economic and social privilege doesn't just give them more to lose from a new economic order; it gives them reason to be more sanguine about the risks of climate change in the first place. This occurred to me as I listened to yet another speaker at the Heartland conference display what can only be described as an utter absence of empathy for the victims of climate disruption. Larry Bell, whose bio describes him as a "space architect," drew plenty of laughs when he told the crowd that a little heat isn't so bad: "I moved to Houston intentionally!" (Houston was, at that time, in the midst of what would turn out to be the state's worst single-year drought on record.) Australian geologist Bob Carter offered that "the world actually does better from our human perspective in warmer times." And Patrick Michaels said people worried about climate change should do what the French did after a devastating 2003 heat wave killed fourteen thousand of their people: "they discovered Walmart and air-conditioning."

Listening to these zingers as an estimated thirteen million

people in the Horn of Africa faced starvation on parched land was deeply unsettling. What makes this callousness possible is the firm belief that if the deniers are wrong about climate change, a few degrees of warming isn't something wealthy people in industrialized countries have to worry about. ("When it rains, we find shelter. When it's hot, we find shade," Texas congressman Joe Barton explained at an energy and environment subcommittee hearing.)

As for everyone else, well, they should stop looking for handouts and busy themselves getting unpoor. When I asked Michaels if rich countries had a responsibility to help poor ones pay for costly adaptations to a warmer climate, he scoffed at the idea, saying that there was no reason to give money to poor countries "because, for some reason, their political system is incapable of adapting." The real solution, he claimed, was more free trade.

This is where the intersection between hard-right ideology and climate denial gets truly dangerous. It's not simply that these "cool dudes" deny climate science because it threatens to upend their dominance-based worldview. It is that their dominance-based worldview provides them with the intellectual tools to write off huge swaths of humanity in the developing world. Recognizing the threat posed by this empathy-exterminating mind-set is a matter of great urgency, because climate change will test our moral character like little before. The US Chamber of Commerce, in its bid to prevent the Environmental Protection Agency from regulating carbon emissions, argued in a petition that in the event of global warming, "populations can acclimatize to warmer climates via a range of behavioral, physiological, and technological adaptations." These adaptations are what I worry about most.

How will we adapt to the people made homeless and jobless by increasingly intense and frequent natural disasters? How will we treat the climate refugees who arrive on our shores in leaky boats? Will we open our borders, recognizing that we created the crisis from which they are fleeing? Or will we build ever more high-tech fortresses and adopt ever more draconian anti-immigration laws? How will we deal with resource scarcity?

We know the answers already. The corporate quest for scarce resources will become more rapacious, more violent. Arable land in Africa will continue to be grabbed to provide food and fuel to wealthier nations. Drought and famine will continue to be used as a pretext to push genetically modified seeds, driving farmers further into debt. We will attempt to transcend peak oil and gas by using increasingly risky technologies to extract the last drops, turning ever-larger swaths of our globe into sacrifice zones. We will fortress our borders and intervene in foreign conflicts over resources, or start those conflicts ourselves. "Free-market climate solutions," as they are called, will be a magnet for speculation, fraud, and crony capitalism, as we are already seeing with carbon trading and the use of forests as carbon offsets. And as climate change begins to affect not just the poor but the wealthy as well, we will increasingly look for techno fixes to turn down the temperature, with massive and unknowable risks.

As the world warms, the reigning ideology that tells us it's everyone for themselves, that victims deserve their fate, that we can master nature, will take us to a very cold place indeed. And it will only get colder, as theories of racial superiority, barely under the surface in parts of the denial movement, make a raging comeback. These theories are not optional: they are necessary to justify the hardening of hearts to the largely blameless victims of climate change in the Global South and in predominately African American cities like New Orleans.

In *The Shock Doctrine* (2007), I explore how the right has systematically used crises, real and trumped up, to push through a brutal ideological agenda designed not to solve the problems that created the crises but, rather, to enrich elites. As the climate crisis begins to bite, it will be no exception. This is entirely predictable. Finding new ways to privatize the commons and to profit from disaster are what our current system is built to do.

The only wild card is whether some countervailing popular movement will step up to provide a viable alternative to this grim future. That means not just an alternative set of policy proposals but an alternative worldview to rival the one at the heart of the ecological crisis—this time, one embedded in interdependence rather than hyperindividualism, reciprocity rather than dominance, and cooperation rather than hierarchy.

Shifting cultural values is, admittedly, a tall order. It calls for the kind of ambitious vision that movements used to fight for a century ago, before everything was broken into single "issues" to be tackled by the appropriate sector of business-minded nongovernmental organizations. Climate change is, in the words of the *Stern Review on the Economics of Climate Change*, "the greatest example of market failure we have ever seen." By all rights, this reality should be filling progressive sails with conviction, breathing new life and urgency into long-standing fights against everything from pro-corporate free trade to financial speculation to industrial agriculture to Third World debt, while elegantly weaving all these struggles into a coherent narrative about how to protect life on earth.

But that isn't happening, at least not so far. It is a painful irony that while the Heartlanders are busily calling climate change a left-wing plot, most leftists have yet to realize that climate science has handed them the most powerful argument against capitalism since William Blake's "dark Satanic Mills" (and, of course, those mills were the beginning of climate change). When demonstrators are cursing out the corruption of their governments and corporate elites in Athens, Madrid, Cairo, Madison, and New York, climate change is often little more than a flesh wound, when it should be the coup de grâce.

Half the problem is that progressives, their hands full with battling systemic economic and racial exclusions, not to mention multiple wars, tend to assume that the big green groups have the climate issue covered. The other half is that many of the biggest big green groups have avoided, with phobic precision, any serious debate on the blindingly obvious roots of the climate crisis: globalization, deregulation, and contemporary capitalism's quest for perpetual growth (the same forces that are responsible for so much destruction in the rest of the economy). The result is that those taking on the failures of capitalism and those fighting for climate action remain two solitudes, with the small but valiant climate justice movement (drawing the connections between racism, inequality, and environmental vulnerability) stringing up a few swaying bridges between them.

The right, meanwhile, has had a free hand to exploit the global economic crisis that began in 2008 to cast climate action as a recipe for economic Armageddon, a surefire way to spike household costs and to block new, much-needed jobs drilling for oil and laying new pipelines. With virtually no loud voices offering a competing vision of how a new economic paradigm could provide a way out of both the economic and ecological crises, this fearmongering has had a ready audience.

Far from learning from past mistakes, a powerful faction in the environmental movement is pushing to go even further down the same disastrous road, arguing that the way to win on climate is to make the cause more palatable to conservative values. This can be heard from the studiously centrist Breakthrough Institute, which is calling for the movement to embrace industrial agriculture and nuclear power instead of agroecological farming and decentralized renewables. It can also be heard from several of the researchers studying the rise in climate denial. Some, like Yale's Kahan, point out that while those who poll as highly "hierarchical" and "individualist" bridle at any mention of regulation, they tend to like big, centralized technologies that confirm their belief that humans can dominate nature. So, he and others argue, environmentalists should start emphasizing responses such as nuclear power and geoengineering (i.e., deliberately intervening in the climate system to counteract global warming) and play up concerns about national security.

The first problem with this strategy is that it doesn't work. For years, big green groups have framed climate action as a way to assert "energy security," while "free-market solutions" are virtually the only ones on the table in the United States. Meanwhile, denialism has soared. The more troubling problem with this approach, however, is that rather than challenging the warped values motivating denialism, it reinforces them. Nuclear power and geoengineering are not solutions to the ecological crisis; they are a doubling down on exactly the kind of short-term hubristic thinking that got us into this mess.

It is not the job of a transformative social movement to reassure members of a panicked, megalomaniacal elite that they are still masters of the universe; nor is it necessary. True, this demographic is massively overrepresented in positions of power, but the solution to that problem is not for the majority of people to change their ideas and values. It is to attempt to change the culture so that this small but disproportionately influential minority, and the reckless worldview it represents, wields significantly less power.

Some in the climate camp are pushing back hard against the appeasement strategy. Tim DeChristopher, who served a two-year jail sentence in Utah for disrupting a compromised auction of oil and gas leases, commented on the right-wing claim that climate action will upend the economy. "I believe we should embrace the charges," he told an interviewer. "No, we are not trying to disrupt the economy, but yes, we do want to turn it upside down. We should not try and hide our vision about what we want to change—of the healthy, just world that we wish to create. We are not looking for small shifts: we want a radical overhaul of our economy and society." He added, "I think once we start talking about it, we will find more allies than we expect."

When DeChristopher articulated this vision for a climate movement fused with one demanding deep economic transformation, it surely sounded to most like a pipe dream. Today, it sounds prophetic. It turns out that a great many people have been hungering for this kind of transformation on many fronts, from the practical to the spiritual.

And new political connections are already being made. The Rainforest Action Network, which has been targeting Bank of America for financing the coal industry, has made common cause with Occupy activists taking aim at the bank over foreclosures. Antifracking activists have pointed out that the same economic model that is blasting the bedrock of the earth to keep the gas flowing is blasting the social bedrock to keep the profits flowing. And then there is the historic movement against the Keystone XL

Pipeline, which this fall has decisively yanked the climate movement out of the lobbyists' offices and into the streets (and jail cells). Anti-Keystone campaigners have noted that anyone concerned about the corporate takeover of democracy need look no further than the corrupt process that led the State Department to conclude that a pipeline carrying dirty tar sands oil across some of the most sensitive land in the country would have "limited adverse environmental impacts." As 350.org's Phil Aroneanu put it, "If Wall Street is occupying President Obama's State Department and the halls of Congress, it's time for the people to occupy Wall Street."

But these connections go beyond a shared critique of corporate power. As Occupiers ask themselves what kind of economy should be built to displace the one crashing all around us, many are finding inspiration in the network of green economic alternatives that has taken root over the past decade—in community-controlled renewable energy projects, in community-supported agriculture and farmers' markets, in economic localization initiatives that have brought main streets back to life, and in the co-op sector.

Not only do these economic models create jobs and revive communities while reducing emissions, but they do so in a way that systematically disperses power—the antithesis of an economy by and for the 1 percent. Omar Freilla, one of the founders of Green Worker Cooperatives in the South Bronx, told me that the experience in direct democracy that thousands are having in plazas and parks as part of movements against economic austerity has been, for many, "like flexing a muscle you didn't know you had." And, he says, now they want more democracy—not just at a meeting but also in their community planning and in their workplaces.

In other words, cultural values are beginning to shift. Today's young organizers are setting out to change policy, but they understand that before that can happen, we have to confront the underlying values of rampant greed and individualism that created the economic crisis. And that begins with embodying, in highly visible ways, radically different ways of treating one another and relating to the natural world.

This deliberate attempt to shift cultural values is not about lifestyle politics; nor is it a distraction from the "real" struggles. Because in the rocky future we have already made inevitable, an unshakable belief in the equal rights of all people and a capacity for deep empathy will be the only things standing between humanity and barbarism. Climate change, by putting us on a firm deadline, can serve as the catalyst for precisely this profound social and ecological transformation.

Culture, after all, is fluid. It can change. It has happened many times in our history. The delegates at the Heartland conference know this, which is why they are so determined to suppress the mountain of evidence proving that their worldview is a threat to life on Earth. The task for the rest of us is to believe, based on that same evidence, that a very different worldview can be our salvation.

GEOENGINEERING: TESTING THE WATERS

Wouldn't it be better to change our behavior—to reduce our use of fossil fuels—before we begin fiddling with the planet's basic life-support systems?

OCTOBER 2012

FOR ALMOST TWENTY YEARS, I'VE BEEN SPENDING TIME ON A CRAGGY STRETCH of British Columbia's shoreline called the Sunshine Coast. A few months ago, I had an experience that reminded me why I love this place, and why I chose to have a child in this sparsely populated part of the world.

It was 5 AM, and my husband and I were up with our three-week-old son. Looking out at the ocean, we spotted two towering, black dorsal fins: orcas, or killer whales. Then two more. We had never seen an orca on this part of the coast, certainly not just a few feet from shore. In our sleep-deprived state, it felt like a miracle, as if the baby had awakened us to make sure we didn't miss this rare visit.

The possibility that the sighting may have resulted from

something less serendipitous did not occur to me until recently, when I read reports of a bizarre ocean experiment off the islands of Haida Gwaii, several hundred miles from where we spotted the orcas swimming.

There, an American entrepreneur named Russ George dumped 120 tons of iron dust off the hull of a rented fishing boat. The plan was to create an algae bloom that would sequester carbon and thereby combat climate change.

George is one of a growing number of would-be geoengineers who advocate high-risk, large-scale technical interventions that would fundamentally change the oceans and skies in order to reduce the effects of global warming. In addition to George's scheme to fertilize the ocean with iron, other geoengineering strategies under consideration include pumping sulfate aerosols into the upper atmosphere to imitate the cooling effects of a major volcanic eruption and "brightening" clouds so they reflect more of the sun's rays back to space.

The risks are huge. Ocean fertilization could trigger dead zones and toxic tides. And multiple simulations have predicted that mimicking the effects of a volcano would interfere with monsoons in Asia and Africa, potentially threatening water and food security for billions of people.

So far, these proposals have mostly served as fodder for computer models and scientific papers. But with George's ocean adventure, geoengineering has decisively escaped the laboratory. If George's account of the mission is to be believed, his actions created an algae bloom in an area half the size of Massachusetts that attracted a huge array of aquatic life from across the region, including whales that could be "counted by the score."

When I read about the whales, I began to wonder: could it be that the orcas I saw swimming north were on their way to feed on

George's bloom? The possibility, unlikely though it is, provides a glimpse into one of the disturbing repercussions of geoengineering: once we start deliberately interfering with the earth's climate systems, whether by dimming the sun or fertilizing the seas, all natural events can begin to take on an unnatural tinge. An absence that might have seemed like a cyclical change in migration patterns or a presence that felt like a miraculous gift suddenly can feel sinister, as if all of nature were being manipulated behind the scenes.

Most news reports characterize George as a "rogue" geoengineer. But what concerns me, after researching the subject for two years, is that far more serious scientists, backed by far deeper pockets, appear poised to actively tamper with the complex and unpredictable natural systems that sustain life on earth—with huge potential for unintended consequences.

In 2010, the chairman of the United States House Committee on Science, Space, and Technology recommended more research into geoengineering; the British government has begun to spend public money in the field. Bill Gates has funneled millions of dollars into geoengineering research.* And he has invested in a company, Intellectual Ventures, that is developing at least two geoengineering tools: the "StratoShield," a nineteen-mile-long hose suspended by helium balloons that would spew sun-blocking sulfur

*Gates is one of the funders of a Harvard University–based research group that has announced it will attempt a groundbreaking field experiment spraying aerosols into the stratosphere in 2019, a plan that has attracted considerable controversy and been delayed several times. According to leading climate scientist Kevin Trenberth, "solar geoengineering is not the answer" to the failure to reduce emissions. "Cutting incoming solar radiation affects the weather and hydrological cycle. It promotes drought. It destabilizes things and could cause wars. The side effects are many and our models are just not good enough to predict the outcomes."

dioxide particles into the sky and a tool that can supposedly blunt the force of hurricanes.

The appeal is easy to understand. Geoengineering offers the tantalizing promise of a climate change fix that would allow us to continue our resource-exhausting way of life, indefinitely. And then there is the fear. Every week seems to bring more terrifying climate news, from reports of ice sheets melting more rapidly than predicted to oceans acidifying far faster than expected. Meanwhile, emissions soar. Is it any wonder that many are pinning their hopes on a break-the-glass-in-case-of-emergency option that scientists have been cooking up in their labs?

But with rogue geoengineers on the loose, it is a good time to pause and ask, collectively, whether we want to go down the geoengineering road. Because the truth is that geoengineering is itself a rogue proposition. By definition, technologies that tamper with ocean and atmospheric chemistry on a planetary scale affect everyone. Yet it is impossible to get anything like unanimous consent for these interventions. Nor could any such consent possibly be informed, given that we don't, and can't, know the full risks involved until these planet-altering technologies are actually deployed.

While the United Nations' climate negotiations proceed from the premise that countries must agree to a joint response to an inherently communal problem, geoengineering raises a very different prospect. For well under a billion dollars, a "coalition of the willing," a single country, or even a wealthy individual could decide to take the climate into their own hands. Jim Thomas of the ETC Group, an environmental watchdog organization, puts the problem like this: "Geoengineering says, 'we'll just do it, and you'll live with the effects.'"

The scariest part is that models suggest that many of the people who could well be most harmed by these technologies are already disproportionately vulnerable to the impacts of climate change. Imagine this: North America decides to send sulfur into the stratosphere to reduce the intensity of the sun, in the hope of saving its corn crops—despite the real possibility of triggering droughts in Asia and Africa. In short, geoengineering would give us (or some of us) the power to exile huge swaths of humanity to sacrifice zones with a virtual flip of the switch.

The geopolitical ramifications are chilling. Climate change is already making it hard to know whether events previously understood as "acts of God" (a freak heat wave in March or a Frankenstorm on Halloween) still belong in that category. But if we start tinkering with the earth's thermostat, deliberately turning our oceans murky green to soak up carbon and bleaching the skies hazy white to deflect the sun, we take our influence to a new level. A drought in India will come to be seen, accurately or not, as a result of a conscious decision by engineers on the other side of the planet to put the region's annual monsoon season in jeopardy. What was once bad luck could come to be seen as a malevolent plot or an imperialist attack.

There will be other, visceral, life-changing consequences. A study published this spring in *Geophysical Research Letters* found that if we inject sulfur aerosols into the stratosphere in order to dial down the sun, the sky would not only become whiter and significantly brighter, but we would also be treated to more intense "volcanic" sunsets. But what kind of relationships can we expect to have with those hyper-real skies? Would they fill us with awe—or with vague unease? Would we feel the same when beautiful wild creatures crossed our paths unexpectedly, as happened to my family this summer? In a popular book on climate change, Bill McKibben warned that we face "The End of Nature." In the age of geoengineering, we might find ourselves confronting the end of miracles, too.

Now that geoengineering threatens to escape the laboratory on a much larger scale than one artificial algae bloom, the real question we face is this: Wouldn't it be better to change our behavior, to reduce our use of fossil fuels, before we begin fiddling with the planet's basic life-support systems?

Because unless we change course, we can expect to hear many more reports about sun shielders and ocean fiddlers like Russ George, whose iron-dumping exploit did more than test a thesis about ocean fertilization. It also tested the waters for future geoengineering experiments. And judging by the muted response so far, the results of George's test are clear: geoengineers proceed, caution be damned.

WHEN SCIENCE SAYS THAT POLITICAL REVOLUTION IS OUR ONLY HOPE

> Most of these scientists were just quietly doing their work measuring ice cores, running global climate models, and studying ocean acidification, only to discover that they "were unwittingly destabilizing the political and social order."

OCTOBER 2013

IN DECEMBER 2012, A PINK-HAIRED COMPLEX SYSTEMS RESEARCHER NAMED Brad Werner made his way through the throng of twenty-four thousand earth and space scientists at the fall meeting of the American Geophysical Union, held annually in San Francisco. This year's conference had some big-name participants, from Ed Stone of NASA's Voyager project, explaining a new milestone on the path to interstellar space, to the filmmaker James Cameron, discussing his adventures in deep-sea submersibles.

But it was Werner's session that was attracting much of the buzz. It was titled "Is Earth F**ked?" (full title: "Is Earth F**ked?

Dynamical Futility of Global Environmental Management and Possibilities for Sustainability via Direct Action Activism").

Standing at the front of the conference room, the geophysicist from the University of California, San Diego, walked the crowd through the advanced computer model he was using to answer that question. He talked about system boundaries, perturbations, dissipation, attractors, bifurcations, and a whole bunch of other stuff largely incomprehensible to those of us uninitiated in complex systems theory. But the bottom line was clear enough: global capitalism had made the depletion of resources so rapid, convenient, and barrier-free that "earth-human systems" were becoming dangerously unstable in response. When pressed by a journalist for a clear answer on the "are we f**ked" question, Werner set the jargon aside and replied, "More or less."

There was one dynamic in the model, however, that offered some hope. Werner termed it "resistance": movements of "people or groups of people" who "adopt a certain set of dynamics that does not fit within the capitalist culture." According to the abstract for his presentation, this includes "environmental direct action, resistance taken from outside the dominant culture, as in protests, blockades and sabotage by indigenous peoples, workers, anarchists and other activist groups."

Serious scientific gatherings don't usually feature calls for mass political resistance, much less direct action and sabotage. But then again, Werner wasn't exactly calling for those things. He was merely observing that mass uprisings of people (along the lines of the abolition movement, the civil rights movement, or Occupy Wall Street) represent the likeliest source of "friction" to slow down an economic machine that is careening out of control. We know that past social movements have "had tremendous influence on . . . how the dominant culture evolved," he pointed out. So, it stands to reason that "if we're thinking about the future of the earth, and the future of our coupling to the environment, we have to include resistance as part of that dynamics." And that, Werner argued, is not a matter of opinion, but "really, a geophysics problem."

Plenty of scientists have been moved by their research findings to take action in the streets. Physicists, astronomers, medical doctors, and biologists have been at the forefront of movements against nuclear weapons, nuclear power, war, and chemical contamination. And in November 2012, *Nature* published a commentary by the financier and environmental philanthropist Jeremy Grantham urging scientists to join this tradition and "be arrested if necessary," because climate change "is not only the crisis of your lives—it is also the crisis of our species' existence."

Some scientists need no convincing. The godfather of modern climate science, James Hansen, is a formidable activist, having been arrested some half-dozen times for resisting mountaintop removal coal mining and tar sands pipelines. (He even left his job at NASA this year in part to have more time for campaigning.) Two years ago, when I was arrested outside the White House at a mass action against the Keystone XL tar sands pipeline, one of the 166 people in cuffs that day was a glaciologist named Jason Box, a world-renowned expert on Greenland's melting ice sheet. "I couldn't maintain my self-respect if I didn't go," Box said at the time, adding that "just voting doesn't seem to be enough in this case. I need to be a citizen also."

This is laudable, but what Werner is doing with his modeling is different. He isn't saying that his research drove him to take action to stop a particular policy; he is saying that his research shows that our entire economic paradigm is a threat to ecological stability.

And, indeed, that challenging this economic paradigm, through mass movement counterpressure, is humanity's best shot at avoiding catastrophe.

That's heavy stuff. But he's not alone. Werner is part of a small but increasingly influential group of scientists whose research into the destabilization of natural systems, particularly the climate system, is leading them to similarly transformative, even revolutionary, conclusions. And for any closet revolutionary who has ever dreamed of replacing the present economic order with one a little less likely to cause Italian pensioners to hang themselves in their homes (as happened recently in the midst of that country's austerity crisis), this work should be of particular interest. Because it makes the ditching of that cruel system in favor of something distinctly fairer no longer a matter of mere ideological preference but, rather, one of species-wide existential necessity.

Leading the pack of these new scientific revolutionaries is one of Britain's top climate experts, Kevin Anderson, the deputy director of the Tyndall Centre for Climate Change Research, which has quickly established itself as one of the United Kingdom's premier climate research institutions. Addressing everyone from the Department for International Development to Manchester City Council, Anderson has spent more than a decade patiently translating the implications of the latest climate science to politicians, economists, and campaigners. In clear and understandable language, he lays out a rigorous road map for emissions reduction, one that provides a decent shot at keeping global temperature rise below the temperature target that most governments have determined would stave off catastrophe.

But in recent years, Anderson's papers and slide shows have become more alarming. Under titles such as "Climate Change: Going Beyond Dangerous . . . Brutal Numbers and Tenuous

Hope," he points out that the chances of staying within anything like safe temperature levels are diminishing fast.

With his colleague Alice Bows, a climate mitigation expert at the Tyndall Centre, Anderson points out that we have lost so much time to political stalling and weak climate policies—all while global consumption (and emissions) ballooned—that we are now facing cuts so drastic that they challenge the fundamental logic of prioritizing GDP growth above all else.

Anderson and Bows inform us that the often-cited long-term mitigation target, an 80 percent emissions cut below 1990 levels by 2050, has been selected purely for reasons of political expediency and has "no scientific basis." That's because climate impacts come not just from what we emit today and tomorrow, but from the cumulative emissions that build up in the atmosphere over time. And they warn that by focusing on targets decades into the future, rather than on what we can do to cut carbon sharply and immediately, there is a serious risk that we will allow our emissions to continue to soar for years to come, thereby blowing through our "carbon budget" and putting ourselves in an impossible position later in the century.

Which is why Anderson and Bows argue that if the governments of developed countries are serious about hitting the agreed-upon international target of keeping warming below 2°C, and if reductions are to respect any kind of equity principle, then the reductions need to be a lot deeper, and they need to come a lot sooner.

Anderson, Bows, and many others warn that 2°C of warming already involves facing an array of hugely damaging climate impacts and that a 1.5°C target would be far safer. Still, to have even a fifty-fifty chance of hitting the 2°C target, the industrialized countries need to start cutting their greenhouse gas emissions by something like 10 percent a year (more if they want to hit 1.5°C), and they need to start right now. But Anderson and Bows go further, pointing out that this target cannot be met with the array of modest carbon pricing or green tech solutions usually advocated by big green groups. These measures will certainly help, to be sure, but they are simply not enough: a 10 percent drop in emissions, year after year, is virtually unprecedented since we started powering our economies with coal. In fact, cuts above 1 percent per year "have historically been associated only with economic recession or upheaval," as the economist Nicholas Stern put it in his 2006 report for the British government.

Even after the Soviet Union collapsed, reductions of this duration and depth did not happen. (The former Soviet countries experienced average annual reductions of roughly 5 percent over a period of ten years.) They did not happen after Wall Street crashed in 2008. (Wealthy countries experienced about a 7 percent emissions drop between 2008 and 2009, but their CO_2 emissions rebounded with gusto in 2010 and emissions in China and India had continued to rise.) Only in the immediate aftermath of the great market crash of 1929 did the United States, for instance, see emissions drop for several consecutive years by more than 10 percent annually, according to historical data from the Carbon Dioxide Information Analysis Center. But that was the worst economic crisis of modern times.

If we are to avoid that kind of carnage while meeting science-based emissions targets, carbon reduction must be managed carefully through what Anderson and Bows describe as "radical and immediate de-growth strategies in the US, EU and other wealthy nations." Which is fine, except that we happen to have an economic system that fetishizes GDP growth above all else, regardless of the human or ecological consequences, and in which the neoliberal political class has utterly abdicated its responsibility to manage anything (since the market is the invisible genius to which everything must be entrusted).

So, what Anderson and Bows are really saying is that there is still time to avoid catastrophic warming, but not within the rules of capitalism as they are currently constructed. Which may be the best argument we have ever had for changing those rules.

In a 2012 essay that appeared in the influential scientific journal *Nature Climate Change*, Anderson and Bows threw down something of a gauntlet, accusing many of their fellow scientists of failing to come clean about the kind of changes that climate change demanded of humanity. On this it is worth quoting the pair at length:

> . . . in developing emission scenarios scientists repeatedly and severely underplay the implications of their analyses. When it comes to avoiding a 2°C rise, "impossible" is translated into "difficult but doable," whereas "urgent and radical" emerge as "challenging"—all to appease the god of economics (or, more precisely, finance). For example, to avoid exceeding the maximum rate of emission reduction dictated by economists, "impossibly" early peaks in emissions are assumed, together with naive notions about "big" engineering and the deployment rates of low-carbon infrastructure. More disturbingly, as emissions budgets dwindle, so geoengineering is increasingly proposed to ensure that the diktat of economists remains unquestioned.

In other words, in order to appear reasonable within neoliberal economic circles, scientists have been dramatically soft-pedaling the implications of their research. By August 2013, Anderson was willing to be even blunter, writing that the boat on gradual change had sailed.

> Perhaps at the time of the 1992 Earth Summit, or even at the turn of the millennium, 2°C levels of mitigation could have been achieved through significant *evolutionary changes within the political and economic hegemony*. But climate change is a cumulative issue! Now, in 2013, we in high-emitting (post-) industrial nations face a very different prospect. Our ongoing and collective carbon profligacy has squandered any opportunity for the "evolutionary change" afforded by our earlier (and larger) 2°C carbon budget. Today, after two decades of bluff and lies, the remaining 2°C budget demands *revolutionary change to the political and economic hegemony* (emphasis theirs).

We probably shouldn't be surprised that some climate scientists are a little spooked by the radical implications of even their own research. Most of them were just quietly doing their work measuring ice cores, running global climate models, and studying ocean acidification, only to discover, as the Australian climate expert and author Clive Hamilton puts it, that they "were unwittingly destabilizing the political and social order."

But there are many people who are well aware of the revolutionary nature of climate science. It's why some of the governments that decided to chuck their climate commitments in favor of digging up more carbon have had to find ever-more-thuggish ways to silence and intimidate their nations' scientists. In Britain, this strategy is becoming more overt, with Ian Boyd, the chief scientific adviser at the Department for Environment, Food and Rural Affairs, writing recently that scientists should avoid "suggesting that policies are either right or wrong" and should express their views "by working with embedded advisers (such as myself), and by being the voice of reason, rather than dissent, in the public arena."

But the truth is getting out anyway. The fact that the

business-as-usual pursuit of profits and growth is destabilizing life on Earth is no longer something we need to read about in scientific journals. The early signs are unfolding before our eyes. And increasing numbers of us are responding accordingly: blockading fracking activity in Balcombe, England; interfering with Arctic drilling preparations in Russian waters (as Greenpeace has done); taking tar sands operators to court for violating Indigenous sovereignty; and countless other acts of resistance large and small. In Brad Werner's computer model, this is the "friction" needed to slow down the forces of destabilization; the great climate campaigner and author Bill McKibben calls it the "antibodies" rising up to fight the planet's "spiking fever."

It's not a revolution yet, but it's a start. And if it spreads, it might just buy us enough time to figure out a way to live on a planet that is distinctly less f**ked.

CLIMATE TIME VS. THE CONSTANT NOW

The climate crisis was hatched in our laps at a moment in history when political and social conditions were uniquely hostile to a problem of this nature and magnitude—that moment being the tail end of the go-go 80s, the blast-off point for the crusade to spread deregulated capitalism around the world.

APRIL 2014

THIS IS A STORY ABOUT BAD TIMING.

One of the most distressing ways that climate change-fueled extinction is already playing out is through what ecologists call "mismatch" or "mistiming." This is the process whereby warming causes animals to fall out of step with a critical food source, particularly at breeding times, when a failure to find enough food can lead to rapid population losses.

The migration patterns of many songbird species, for instance, have evolved over millennia so that eggs hatch precisely when food sources such as caterpillars are at their most abundant, providing parents with ample nourishment for their hungry young. But because spring now often arrives early, the caterpillars are hatching earlier, too, which means that in some areas they are less plentiful when the chicks hatch, with a number of possible long-term impacts on survival.

Similarly, in West Greenland, caribou are arriving at their calving grounds only to find themselves out of sync with the forage plants they have relied on for thousands of years, now growing earlier thanks to rising temperatures. That is leaving female caribou with less energy for lactation and reproduction, a mismatch that has been linked to sharp decreases in calf births and survival rates.

Scientists are studying cases of climate-related mistiming among dozens of species, from Arctic terns to pied flycatchers. But there is one important species they are missing: us. *Homo sapiens*. We, too, are suffering from a terrible case of climate-related mistiming, albeit in a cultural-historical, rather than a biological, sense. Our problem is that the climate crisis was hatched in our laps at a moment in history when political and social conditions were uniquely hostile to a problem of this nature and magnitude—that moment being the tail end of the go-go 80s, the blast-off point for the crusade to spread deregulated capitalism around the world. Climate change is a collective problem demanding collective action on a scale that humanity has never actually accomplished. Yet it entered mainstream consciousness in the midst of an ideological war being waged on the very idea of the collective sphere.

This deeply unfortunate mistiming has created all sorts of barriers to our ability to respond effectively to this crisis. It has meant that corporate power was ascendant at the very moment when we needed to exert unprecedented controls over corporate behavior in order to protect life on Earth. It has meant that *regulation* was a dirty word just when we needed those powers most. It has meant that we are ruled by a class of politicians who know only how to dismantle and starve public institutions just when they most need to be fortified and reimagined. And it has meant that we are saddled with an apparatus of "free-trade" deals that tie the hands of policymakers just when they need maximum flexibility to achieve a massive energy transition.

Confronting these various structural barriers to the next economy, and articulating a captivating vision for that postcarbon way of life, is the critical work of any serious climate movement. But it's not the only task at hand. We also have to confront how the mismatch between climate change and market domination has created barriers within our very selves, making it harder for us to look at this most pressing of humanitarian crises with anything more than furtive, terrified glances. Because of the way our daily lives have been altered by both market and technological triumphalism, we lack many of the observational tools necessary to convince ourselves that climate change is indeed an emergency—let alone the confidence to believe that a different way of living is possible.

And little wonder: just when we needed to gather, our public sphere was disintegrating; just when we needed to consume less, consumerism took over virtually every aspect of our lives; just when we needed to slow down and notice, we sped up; and just when we needed longer time horizons, we were able to see only the immediate present, trapped in the forever now of our constantly refreshed social media feeds.

This is our climate change mismatch, and it affects not just our species but potentially every other species on the planet as well.

The good news is that unlike reindeer and songbirds, we humans are blessed with the capacity for advanced reasoning and therefore the ability to adapt more deliberately—to change old patterns of behavior with remarkable speed. If the ideas that rule our culture are stopping us from saving ourselves, then it is within our power to change those ideas. But before that can happen, we first need to understand the nature of our personal climate mismatch.

BEING CONSUMERS IS ALL WE KNOW

Climate change demands that we consume less, but being consumers is all we know. Climate change is not a problem that can be solved simply by changing what we buy—a hybrid instead of an SUV, some carbon offsets when we get on a plane. At its core, it is a crisis born of overconsumption by the comparatively wealthy, which means the world's most manic consumers are going to have to consume less so that others can have enough to live.

The problem is not "human nature," as we are so often told. We weren't born having to shop this much, and we have, in our recent past, been just as happy (in many cases, happier) consuming significantly less. The problem is the inflated role that consumption has come to play in our particular era.

Late capitalism teaches us to create ourselves through our consumer choices: shopping is how we form our identities, find community, and express ourselves. Thus, telling people that they can't shop as much as they want to because the planet's support systems are overburdened can be understood as a kind of attack, akin to telling them that they cannot truly be themselves. This is likely why, of environmentalism's original "three *R*s" (reduce, reuse, recycle), only the third has ever gotten any traction, since it allows us to keep on shopping as long as we put the refuse in the right box.*

*We now know that much of that third *R* has been for naught: in cities across North America, mountains of plastic takeout containers and junk mail that consumers thought were headed to recycling depots to be turned into more

The other two, which require that we consume less, were pretty much dead on arrival.

CLIMATE CHANGE IS SLOW, BUT WE ARE FAST

When you are racing through a rural landscape on a bullet train, it looks as if everything you are passing were standing still: people, tractors, cars on country roads. They aren't, of course. They are moving, but at a speed so slow compared with the train that they appear static.

So it is with climate change. Our culture, powered by fossil fuels, is that bullet train, hurtling forward toward the next quarterly report, the next election cycle, the next bit of diversion or piece of personal validation via our smartphones and tablets. Our changing climate is like the landscape outside the window: from our racy vantage point it can appear static, but it is moving, its slow progress measured in receding ice sheets, swelling waters, and incremental temperature rises. If left unchecked, climate change will most certainly speed up enough to capture our fractured attention—island nations wiped off the map and city-drowning superstorms tend to do that. But, by then, it may be too late for our actions to make a difference because the era of tipping points will likely have begun.

useful items are actually going straight into landfills or being incinerated, both powerful sources of greenhouse gases. This is because China, in 2018, severely reduced the amount of recyclable waste it was willing to accept after the low-margin industry was found to have caused severe health and environmental impacts.

CLIMATE CHANGE IS PLACE-BASED, BUT WE ARE EVERYWHERE AT ONCE

The problem is not just that we are moving too quickly. It is also that the terrain on which climate changes are taking place is intensely local: an early blooming of a particular flower, an unusually thin layer of ice on a lake, sap failing to flow in a maple tree, the late arrival of a migratory bird. Noticing those kinds of subtle changes requires an intimate connection to a specific ecosystem. That kind of communion happens only when we know a place deeply, not just as scenery but also as sustenance, and when local knowledge is passed on with a sense of sacred duty from one generation to the next.

But that is increasingly rare in the urbanized, industrialized world. Few of us live where our ancestors are buried. Many of us abandon our homes lightly—for a new job, a new school, a new love. And as we do so, we are severed from whatever knowledge of place we managed to accumulate at the previous stop, and from the knowledge amassed by our ancestors (who, in my case like that of so many others, migrated repeatedly themselves).

Even for those of us who manage to stay put, daily existence is increasingly disconnected from the physical places where we reside. We live much of our lives through the portals of screens and navigate the physical world not with our senses but with miniature maps on our phones.

Shielded from the elements as we are in our climate-controlled homes, workplaces, and cars, we can find the changes unfolding in the natural world passing us by. We might have no idea that a historic drought is destroying the crops on the farms that surround our urban homes, given that the supermarkets still display miniature mountains of imported produce, with more coming in by truck all day. It takes something huge—a hurricane that passes all previous high-water marks, or a flood destroying thousands of homes—for us to notice that something is truly amiss. And even then, we have trouble holding on to that knowledge for long, as we are quickly ushered along to the next crisis before these truths have a chance to sink in.*

Climate change, meanwhile, is busily adding to the ranks of the rootless every day, as natural disasters, failed crops, starving livestock, and climate-fueled ethnic conflicts force more and more people to leave their ancestral homes. And with every human migration, more crucial connections to specific places are lost, leaving yet fewer people with the tools to listen closely to the land.

OUT OF SIGHT IS OUT OF OUR MINDS

Climate pollutants are invisible, and many of us have stopped believing in what we cannot see. When former BP chief executive Tony Hayward told reporters that we shouldn't worry much about the oil and chemical dispersants gushing into the Gulf of Mexico after the Deepwater Horizon disaster because it "is a very big ocean," he was merely voicing one of our culture's most cherished beliefs: that what we can't see won't hurt us and, indeed, barely exists.

So much of our economy relies on the assumption that there is always an "away" into which we can throw our waste. There's the

*I wrote to check in on a friend in Los Angeles after a wall of fire had encroached on the city like a beast from the Apocalypse. "There were a few days where the sky was so thick it tasted like secondhand smoke in an '80s nightclub, and all anyone was talking about was evacuation plans," she reported. "But now everyone has gone back to business as usual and it makes me wonder what exactly it will take for people to . . . not do that." What indeed?

away where our garbage goes when it is taken from the curb, and the away where our waste goes when it is flushed down the drain. There's the away where the minerals and metals that make up our goods are extracted, and the away where those raw materials are turned into finished products. But the lesson of the BP spill, in the words of ecological theorist Timothy Morton, is that ours is "a world in which there is no 'away.'"

When I published *No Logo* at the turn of this century, readers were shocked to discover the abusive conditions under which their clothing and gadgets were manufactured. But most of us have since learned to live with it—not to condone it, exactly, but to be in a state of constant forgetfulness about the real-world costs of our consumption. The "aways" of those factories has largely faded back into oblivion.

This is one of the ironies of being told that we live in a time of unprecedented connection. It is true that we can and do communicate across vast geographies with an ease and speed that were unimaginable only a generation ago. But in the midst of this global web of chatter, we somehow manage to be *less* connected to the people with whom we are most intimately enmeshed: the young women in Bangladesh's firetrap factories who make the clothes on our bodies, or the children in the Democratic Republic of the Congo whose lungs are filled with dust from mining cobalt for the phones that have become extensions of our arms. Ours is an economy of ghosts, of deliberate blindness.

Air is the ultimate unseen, and the greenhouse gases that warm it are our most elusive ghosts of all. Philosopher David Abram points out that for most of human history, it was precisely this unseen quality that gave the air its power and commanded our respect. "Called Sila, the wind-mind of the world, by the Inuit; Nilch'i, or Holy Wind, by the Navajo; Ruach, or rushing-spirit, by the ancient Hebrews," the atmosphere was "the most mysterious and sacred dimension of life."

But in our time "we rarely acknowledge the atmosphere as it swirls between two persons." Having forgotten the air, Abram writes, we have made it our sewer, "the perfect dump site for the unwanted by-products of our industries . . . Even the most opaque, acrid smoke billowing out of the pipes will dissipate and disperse, always and ultimately dissolving into the invisible. It's gone. Out of sight, out of mind."

THE TIME FRAMES THAT ESCAPE US

Another part of what makes climate change so very difficult for many of us to grasp is that we live in a culture of the perpetual present, one that deliberately severs itself from the past that created us and the future we are shaping with our actions. Climate change is about how what we did generations in the past will inescapably affect not just the present, but generations in the future. These time frames are a language that has become foreign to most of us in our digitized times.

This is not about passing individual judgment, nor about berating ourselves for our shallowness, rootlessness, or the shredded state of our attention spans. Rather, it is about recognizing that most of us living in urban centers and wealthy countries are products of an industrial project, one intimately and historically linked to fossil fuels and then taken supernova by digital tech.

And just as we have changed before, we can change again. After listening to the great farmer-poet Wendell Berry deliver a lecture on how we each have a duty to love our "homeplace" more than any other, I asked him if he had any advice for rootless people like me and my friends, who disappear into our screens and always seem to be shopping for the perfect community where we should put down roots. "Stop somewhere," he replied. "And begin the thousand-year-long process of knowing that place."

That's good advice on lots of levels, because in order to win this fight of our lives, we all need a place to stand.

STOP TRYING TO SAVE THE WORLD ALL BY YOURSELF

The very idea that we, as atomized individuals, could play a significant part in stabilizing the planet's climate is objectively nuts.

JUNE 2015
COLLEGE OF THE ATLANTIC COMMENCEMENT ADDRESS

USUALLY, A COMMENCEMENT ADDRESS TRIES TO EQUIP GRADUATES WITH A moral compass for their post-university life. You hear stories that end with clear lessons like "Money can't buy happiness," "Be kind," "Don't be afraid to fail."

But my sense is that very few of you are flailing around trying to sort out right from wrong. Quite remarkably, you knew you wanted to go not just to an excellent college, but to an excellent socially and ecologically engaged college. A school surrounded by biological diversity and suffused with tremendous human diversity, with a student population that spans the globe. You also knew that strong community mattered more than almost anything. That's more self-awareness and self-direction than most people have when they leave graduate school—and somehow you had it when you were still in high school.

Which is why I am going to skip the homilies and get down to business: the historical moment into which you graduate—with climate change, wealth concentration, and racialized violence all reaching breaking points.

How do we help most? How do we best serve this broken world? We know that time is short, especially when it comes to climate change. We all hear the clock ticking loudly in the background.

But that doesn't mean that climate change trumps everything else. It means we need to create integrated solutions, ones that radically bring down emissions while tackling structural inequality *and* making life tangibly better for the majority. This is no pipe dream; we have living examples from which to learn. Germany's energy transition has created four hundred thousand jobs in renewables in just over a decade, and not just cleaned up energy but made it fairer, so that many energy grids are owned and controlled by hundreds of cities, towns, and cooperatives. They still have a long way to go in phasing out coal, but they have now started in earnest. New York City just announced a climate plan that, if enacted, would bring eight hundred thousand people out of poverty by 2025 by investing massively in transit and affordable housing and raising the minimum wage.

The holistic leap we need is within our grasp. And know that there is no better preparation for that grand project than your deeply interdisciplinary education in human ecology. You were made for this moment. No, that's not quite right: you somehow knew to make yourselves for this moment.

But much rests on the choices we all make in the next few years. "Don't be afraid to fail" may be a standard commencement address life lesson. Yet it doesn't work for those of us who are part of the climate justice movement, where being afraid of failure is perfectly rational.

Because, let's face it: The generations before you used up more than just your share of atmospheric space. We used up your share of big failures, too—perhaps the ultimate intergenerational injustice. That doesn't mean that we all can't still make mistakes. We can and we will. But Alicia Garza, one of the inspiring founders of Black Lives Matter, talks about how we have to "make new mistakes."

Sit with that one for a minute. Let's stop making the same old mistakes. Here are a few, but I trust that you will silently add your own: Projecting messianic fantasies onto politicians. Thinking the market will fix it. Building a movement made up entirely of upper-middle-class white people and then wondering why people of color don't want to join "our movement." Tearing each other to bloody shreds because it's easier to do that than go after the forces most responsible for this mess. These are social change clichés, and they are getting really boring.

We don't have the right to demand perfection from each other. But we do have the right to expect progress. To demand evolution. So, let's make some new mistakes. Let's make new mistakes as we break through our silos and build the kind of beautifully diverse and justice-hungry movement that actually has a chance of winning—winning against the powerful interests that want us to keep failing.

With this in mind, I want to talk about an old mistake that I see reemerging. It has to do with the idea that since attempts at big systemic change have failed, all we can do is act small. Some of you will relate. Some of you won't. But I suspect all of you will have to deal with this tension in your future work.

A story: When I was twenty-six, I went to Indonesia and the Philippines to do research for my first book, *No Logo*. I had a simple goal: to meet the workers making the clothes and electronics that my friends and I purchased. And I did. I spent evenings on concrete floors in squalid dorm rooms where teenage girls, sweet and giggly, spent their scarce nonworking hours. Eight or even ten to a room. They told me stories about not being able to leave their machines to pee. About bosses who hit and harassed. About not having enough money to buy dried fish to go with their rice.

They knew they were being badly exploited, that the garments and gadgets they were making were being sold for more than they would make in a month. One seventeen-year-old said to me, "We make computers, but we don't know how to use them."

So, one thing I found slightly jarring was that some of these same workers wore clothing festooned with knockoff trademarks of the very multinationals that were responsible for these conditions: Disney characters or Nike check marks. At one point, I asked a local labor organizer about this. Wasn't it strange—a contradiction?

It took a very long time for him to understand the question. When he finally did, he looked at me with something like pity. You see, for him and his colleagues, individual consumption wasn't considered to be in the realm of politics at all. Power rested not in what you did as one person, but what you did as many people, as one part of a large, organized, and focused movement. For him, this meant organizing workers to go on strike for better conditions, and eventually it meant winning the right to unionize. What you ate for lunch or happened to be wearing was of absolutely no concern whatsoever.

This was striking to me, because it was the mirror opposite of my culture back home in Canada. Where I came from, you expressed your political beliefs, first and very often last, through personal lifestyle choices. By loudly proclaiming your vegetarianism. By shopping fair trade and local, and boycotting big, evil brands.

These very different understandings of social change came up again and again a couple of years later, once my book came out. I would give talks about the need for international protections for the right to unionize. About the need to rewire our global trading system so it didn't encourage a race to the bottom. And yet, at the end of those talks, the first question from the audience reliably was "What kind of sneakers are okay to buy?" "What brands are ethical?" "Where do you buy your clothes?" "What can I do, as an individual, to change the world?"

Fifteen years after I published *No Logo*, I still find myself facing very similar questions. These days, I give talks about how the same economic model that superpowered multinationals to seek out cheap labor in Indonesia and China also supercharged global greenhouse gas emissions. And, invariably, the hand goes up: "Tell me what I can do as an individual." Or maybe "as a business owner."

The hard truth is that the answer to the question "What can I, as an individual, do to stop climate change?" is: nothing. You can't do anything. In fact, the very idea that we, as atomized individuals, even lots of atomized individuals, could play a significant part in stabilizing the planet's climate system or changing the global economy is objectively nuts. We can only meet this tremendous challenge together, as part of a massive and organized global movement.

The irony is that people with relatively little power tend to understand this far better than those with a great deal more power. The workers I met in Indonesia and the Philippines knew all too well that governments and corporations did not value their voice or even their lives as individuals. And because of this, they were driven to act not only together, but on a rather large political canvas. To try to change the policies in factories that employ thousands of workers, or in export zones that employ tens of thousands. Or the labor laws in an entire country of millions. Their sense of individual powerlessness pushed them to be politically ambitious, to demand structural changes.

In contrast, here in wealthy countries, we are told how powerful we are as individuals all the time. As consumers. Even individual activists. And the result is that despite our power and privilege, we often end up acting on canvases that are unnecessarily small—the canvas of our own lifestyle, or maybe our neighborhood or town. Meanwhile, we abandon the structural changes, the policy and legal work, to others.

This is not to belittle local activism. Local is critical. Local organizing is winning big fights against fracking and oil pipelines. Local is showing us what the postcarbon economy looks and feels like.

And small examples inspire bigger ones. College of the Atlantic was one of the first schools to divest from fossil fuels. And you made the decision, I am told, in a week. It took that kind of leadership from small schools that knew their values to push more, shall we say, insecure institutions to follow suit. Like Stanford University. Like Oxford University. Like the British royal family. Like the Rockefeller family. All have joined the movement since you did. So, local matters, but local is not enough.

I got a vivid reminder of this when I visited Red Hook, Brooklyn, in the immediate aftermath of Superstorm Sandy. Red Hook was one of the hardest-hit neighborhoods and is home to an amazing community farm, a place that teaches kids from nearby housing projects how to grow healthy food, provides composting for a huge number of residents, hosts a weekly farmers' market, and runs a terrific community-supported agriculture program. In short, it was doing everything right: reducing food miles, staying away from petroleum inputs, sequestering carbon in the soil, reducing landfill by composting, fighting inequality and food insecurity.

But when the storm came, none of that mattered. The entire harvest was lost, and the fear was the storm water would make the soil toxic. They could buy new soil and start over, but the farmers I met there knew that unless other people were out there fighting to lower emissions on a systemic and global level, then this kind of loss would occur again and again.

It's not that one sphere is more important than the other. It's that we have to do both: the local and the global. The resistance and the alternatives. The "nos" to what we cannot survive and the "yeses" that we need to thrive.

Before I leave you, I want to stress one other thing. And please listen, because it's important. It is true that we have to do it all. That we have to change everything. But you personally do not have to do everything. This is not all on you.

One of the real dangers of being brilliant, sensitive young people who hear the climate clock ticking loudly is the danger of taking on too much. Which is another manifestation of that inflated sense of our own importance.

It can seem that every single life decision—whether to work at a national NGO or a local permaculture project or a green startup; whether to work with animals or with people; whether to be a scientist or an artist; whether to go to grad school or have kids—carries the weight of the world.

I was struck by this impossible burden some of you are placing on yourselves when I was contacted recently by a twenty-one-year-old Australian science student named Zoe Buckley Lennox. At the time she reached me, she was camped out on top of Shell's Arctic drilling rig in the middle of the Pacific. She was one of six Greenpeace activists who had scaled the giant rig to try to slow its passage and draw attention to the insanity of drilling for oil in the Arctic. They lived up there in the howling winds for a week.

While they were still up there, I arranged to call Zoe on the Greenpeace satellite phone, just to personally thank her for her courage. Do you know what she did? She asked me, "How do you know you are doing the right thing? I mean, there is divestment. There is lobbying. There's the Paris climate conference."

And I was touched by her seriousness, but I also wanted to weep. Here she was, doing one of the more incredible things imaginable—freezing her butt off trying to physically stop Arctic drilling with her body. And up there in her seven layers of clothing and climbing gear, she was still beating herself up, wondering whether she should be doing something else.

What I told her is what I will tell you. What you are doing is amazing. And what you do next will be amazing, too. Because you are not alone. You are part of a movement. And that movement is organizing at the United Nations and running for office and getting their schools to divest and trying to block Arctic drilling in Congress and the courts. And on the open water. All at the same time.

And, yes, we need to grow faster and do more. But the weight of the world is not on any one person's shoulders: Not yours. Not Zoe's. Not mine. It rests in the strength of the project of transformation that millions are already a part of.

That means we are free to do the kind of work that will sustain us, so that we can all stay in this movement for the long run. Because that's what it will take.

A RADICAL VATICAN?

> People of faith, particularly missionary faiths, believe deeply in something that a lot of secular people aren't so sure about: that all human beings are capable of profound change. . . . That, after all, is the essence of conversion.

JUNE 29, 2015—PACKING

WHEN I WAS FIRST ASKED TO SPEAK AT A VATICAN PRESS CONFERENCE ON POPE Francis's recently published climate change encyclical, "Laudato sì," I was convinced that the invitation would soon be rescinded. Now the press conference and, after it, a two-day symposium to explore the encyclical is just two days away. This is actually happening.

As usual ahead of stressful trips, I displace all my anxiety onto wardrobe. The forecast for Rome in the first week of July is punishingly hot, up to 95°F (35°C). Women visiting the Vatican are supposed to dress modestly, no exposed legs or upper arms. Long, loose cottons are the obvious choice, the only problem being that I have a deep-seated sartorial aversion to anything with the whiff of hippie.

Surely the Vatican press room has air-conditioning. Then

again, "Laudato sì" makes a point of singling it out as one of many "harmful habits of consumption which, rather than decreasing, appear to be growing all the more." Will the powers that be make a point of ditching the climate control just for this press conference? Or will they keep it on and embrace contradiction, as I am doing by supporting the Pope's bold writings on how responding to the climate crisis requires deep changes to our growth-driven economic model, while disagreeing with him about a whole lot else?

To remind myself why this is worth all the trouble, I reread a few passages from the encyclical. In addition to laying out the reality of climate change, it spends considerable time exploring how the culture of late capitalism makes it uniquely difficult to address, or even focus upon, this civilizational challenge. "Nature is filled with words of love," Francis writes, "but how can we listen to them amid constant noise, interminable and nerve-wracking distractions, or the cult of appearances?"

I glance shamefully around at the strewn contents of my closet. (Look: some of us don't get to wear the same white getup everywhere . . .)

JULY 1—THE F-WORD

Four of us are scheduled to speak at the Vatican press conference, including one of the chairs of the UN Intergovernmental Panel on Climate Change. All except me are Catholic. In his introduction, Father Federico Lombardi, the director of the Holy See Press Office, describes me as a "secular Jewish feminist," a term I used in my prepared remarks but never expected him to repeat. Everything else Father Lombardi says is in Italian, but these three words are spoken slowly and in English, as if to emphasize their foreignness.

The first question directed my way is from Rosie Scammell,

with the Religion News Service: "I was wondering how you would respond to Catholics who are concerned by your involvement here, and other people who don't agree with certain Catholic teachings?"

This is a reference to the fact that some traditionalists have been griping about all the heathens, including UN secretary-general Ban Ki-moon and a roster of climate scientists, who were spotted inside these ancient walls in the run-up to the encyclical's publication. The fear is that discussion of planetary overburden will lead to a weakening of the Church's opposition to birth control and abortion. As the editor of a popular Italian Catholic website put it recently, "The road the church is heading down is precisely this: To quietly approve population control while talking about something else."

I respond that I am not here to broker a merger between the secular climate movement and the Vatican. However, if Pope Francis is correct that responding to the climate crisis requires fundamental changes to our economic model—and I think he is correct—then it will take an extraordinarily broad-based movement to demand those changes, one capable of navigating political disagreements.

After the press conference, a journalist from the United States tells me that she has "been covering the Vatican for twenty years, and I never thought I would hear the word 'feminist' from that stage."

The air-conditioning, for the record, was left on.

The British and Dutch ambassadors to the Holy See host a dinner for the conference's organizers and speakers. Over wine and grilled salmon, discussion turns to the political ramifications of the Pope's upcoming trip to the United States. One of the guests most preoccupied with this subject is from an influential American Catholic organization. "The Holy Father isn't making it easy for us by going to Cuba first," he says.

I ask him how spreading the message of "Laudato sì" is going back home. "The timing was bad," he says. "It came out around the same time as the Supreme Court ruling on gay marriage, and that kind of sucked all the oxygen out of the room." That's certainly true. Many US bishops welcomed the encyclical, but not with anything like the Catholic firepower expended to denounce the Supreme Court decision a week later.

The contrast is a vivid reminder of just how far Pope Francis has to go in realizing his vision of a Church that spends less time condemning people over abortion, contraception, and whom they marry, and more time fighting for the trampled victims of a highly unequal and unjust economic system. When climate justice had to fight for airtime with denunciations of gay marriage, it didn't stand a chance.

On the way back to the hotel, looking up at the illuminated columns and dome of St. Peter's Basilica, it strikes me that this battle of wills may be the real reason such eclectic outsiders are being invited inside this cloistered world. We're here because many powerful Church insiders simply cannot be counted upon to champion Francis's transformative climate message—and some would clearly be happy to see it buried alongside the many other secrets entombed in this walled enclave.

Before bed, I spend a little more time with "Laudato sì," and something jumps out at me. In the opening paragraph, Pope Francis writes that "our common home is like a sister with whom we share our life and a beautiful mother who opens her arms to embrace us." He quotes Saint Francis of Assisi's "Canticle of the Creatures," which states, "Praise be to you, my Lord, through our Sister, Mother Earth, who sustains and governs us, and who produces various fruit with colored flowers and herbs."

Several paragraphs down, the encyclical notes that Saint

Francis had "communed with all creation, even preaching to the flowers, inviting them 'to praise the Lord, just as if they were endowed with reason.'" According to Saint Bonaventure, the encyclical says, the thirteenth-century friar "would call creatures, no matter how small, by the name of 'brother' or 'sister.'"

Later in the text, pointing to various biblical directives to care for animals that provide food and labor, Pope Francis comes to the conclusion that "the Bible has no place for a tyrannical anthropocentrism unconcerned for other creatures."

Challenging anthropocentrism is ho-hum stuff for ecologists, but it's something else for the pinnacle of the Catholic Church. You don't get much more human-centered than the persistent Judeo-Christian interpretation that God created the entire world specifically to serve Adam's every need. As for the idea that we are part of a family with all other living beings, with the earth as our life-giving mother, that, too, is familiar to eco-ears. But from the Church? Replacing a maternal Earth with a Father God and draining the natural world of its sacred power were what stamping out paganism, animism, and pantheism were all about.

By asserting that nature has a value in and of itself, Francis is overturning centuries of theological interpretation that regarded the natural world with outright hostility—as a misery to be transcended and an "allurement" to be resisted. Of course, there have been parts of Christianity that stressed that nature was something valuable to steward and protect—some even celebrated it—but mostly as a set of resources to sustain humans.

Francis is not the first pope to express deep environmental concern—John Paul II and Benedict XVI did as well. But those popes didn't tend to call the earth our "sister, mother" or assert that chipmunks and trout are our siblings.

JULY 2—BACK FROM THE WILDERNESS

In St. Peter's Square, the souvenir shops are selling Pope Francis mugs, calendars, aprons—and stacks and stacks of bound copies of "Laudato sì," available in multiple languages. Window banners advertise its presence. At a glance, it looks like just another piece of papal schlock, not a document that could transform Church doctrine.

This morning is the opening of "People and Planet First: The Imperative to Change Course," a two-day gathering to shape an action plan around "Laudato sì," organized by the International Alliance of Catholic Development Agencies and the Pontifical Council for Justice and Peace. Speakers include Mary Robinson, the former president of Ireland and a current UN Special Envoy for Climate Change, and Enele Sopoaga, the prime minister of Tuvalu, an island nation whose existence is under threat from rising seas.

A soft-spoken bishop from Bangladesh leads an opening prayer, and Cardinal Peter Kodwo Appiah Turkson, a major force behind the encyclical, delivers the first keynote. At sixty-six, Turkson has gone gray at the temples, but his round cheeks are still youthful. Many speculate that this could be the man to succeed the seventy-eight-year-old Francis, becoming the first African pope.

Most of Turkson's talk is devoted to citing earlier papal encyclicals as precedents for "Laudato sì." His message is clear: this is not about one pope; it's part of a Catholic tradition of seeing the earth as a sacrament and recognizing a "covenant" (not a mere connection) between human beings and nature.

At the same time, the cardinal points out that "the word *stewardship* only appears twice" in the encyclical. The word *care*, on the other hand, appears dozens of times. This is no accident, we are told. While stewardship speaks to a relationship based on duty, "when one cares for something it is something one does with passion and love."

This passion for the natural world is part of what has come to be called "the Francis factor," and it clearly flows from a shift in geographic power within the Catholic Church. Francis is from Argentina, and Turkson from Ghana. One of the most vivid passages in the encyclical—"Who turned the wonderworld of the seas into underwater cemeteries bereft of color and life?"—is a quotation from a statement of the Catholic Bishops' Conference of the Philippines.

This reflects the reality that in large parts of the Global South, the more anthropocentric elements of Christian doctrine never entirely took hold. Particularly in Latin America, with its large indigenous populations, Catholicism wasn't able to fully displace cosmologies that centered on a living and sacred Earth, and the result was often a Church that fused Christian and indigenous worldviews. With "Laudato sì," that fusion has finally reached the highest echelons of the Church.

Yet Turkson seems to gently warn the crowd here not to get carried away. Some African cultures "deified" nature, he says, but that is not the same as "care." The earth may be a mother, but God is still the boss. Animals may be our relatives, but humans are not animals. Still, once an official papal teaching challenges something as central as human dominion over the earth, is it really possible to control what will happen next?

This point is made forcefully by the Irish-Catholic priest and theologian Seán McDonagh, who was part of the drafting process for the encyclical. His voice booming from the audience, he urges us not to hide from the fact that the love of nature embedded in the encyclical represents a profound and radical shift from traditional Catholicism. "We are moving to a new theology," he declares.

To prove it, he translates a Latin prayer that was once commonly recited after Communion during the season of Advent. "Teach us to despise the things of the earth and to love the things of heaven." Overcoming centuries of loathing the corporeal world is no small task, and, McDonagh argues, it serves little purpose to downplay the work ahead.

It's thrilling to witness such radical theological challenges being batted around inside the curved wooden walls of an auditorium named after Saint Augustine, the theologian whose skepticism of things bodily and material so profoundly shaped the Church. But I would imagine that for the conspicuously silent men in black robes in the front row, who study and teach in this building, it is also a little terrifying.

This evening's dinner is much more informal: a sidewalk trattoria with a handful of Franciscans from Brazil and the United States, as well as McDonagh, who is treated by the others as an honorary member of the order.

My dinner companions have been some of the biggest troublemakers within the Church for years, the ones taking Christ's protosocialist teachings seriously. Patrick Carolan, the Washington, DC–based executive director of the Franciscan Action Network, is one of them. Smiling broadly, he tells me that, at the end of his life, Vladimir Lenin supposedly said that what the Russian Revolution had really needed was not more Bolsheviks but ten Saint Francises of Assisi.

Now, all of a sudden, these outsiders share many of their views with the most powerful Catholic in the world, the leader of a flock of 1.2 billion people. Not only did this pope surprise everyone by calling himself Francis, as no pope ever had before him, but he appears to be determined to revive the most radical Franciscan teachings. Moema de Miranda, a powerful Brazilian social leader, who was wearing a wooden Franciscan cross, says that it feels "as if we are finally being heard."

For McDonagh, the changes at the Vatican are even more striking. "The last time I had a papal audience was 1963," he tells me over *spaghetti alle vongole*. "I let three popes go by." And yet here he is, back in Rome, having helped draft the most talked-about encyclical anyone can remember.

McDonagh points out that it's not just Latin Americans who figured out how to reconcile a Christian God with a mystical Earth. The Irish Celtic tradition also managed to maintain a sense of "divine in the natural world. Water sources had a divinity about them. Trees had a divinity to them." But in much of the rest of the Catholic world, all this was wiped out. "We are presenting things as if there is continuity, but there wasn't continuity. That theology was functionally lost." (It's a sleight of hand that many conservatives are noticing. POPE FRANCIS, THE EARTH IS NOT MY SISTER, reads a recent headline in *The Federalist*, a right-wing online magazine.)

As for McDonagh, he is thrilled with the encyclical, although he wishes it had gone even further in challenging the idea that the earth was created as a gift to humans. How could that be so when we know it was here billions of years before we arrived?

I ask how the Bible could survive this many fundamental challenges—doesn't it all fall apart at some point? He shrugs, telling me that scripture is ever evolving, and should be interpreted in historical context. If Genesis needs a prequel, that's not such a big deal. Indeed, I get the distinct sense that he'd be happy to be part of the drafting committee.

JULY 3—CHURCH, EVANGELIZE THYSELF

I wake up thinking about stamina. Why did Franciscans such as Patrick Carolan and Moema de Miranda stick it out for so long in an institution that didn't reflect many of their deepest beliefs and values—only to live to see a sudden shift that many here can explain only with allusions to the supernatural? Carolan shared with me that he had been abused by a priest at age twelve. He is enraged by the cover-ups, and yet he did not let it drive him permanently from his faith. What kept them there?

I put this to Miranda when I see her at the end of Mary Robinson's lecture. (Robinson had gently criticized the encyclical for failing to adequately emphasize the role of women and girls in human development.)

Miranda corrects me, saying that she is not actually one of those who stuck it out for much of their lifetimes. "I was an atheist for years and years, a Communist, a Maoist. Until I was thirty-three. And then I was converted." She described it as a moment of pure realization: "Wow, God exists. And everything changed."

I asked her what precipitated this, and she hesitates, and laughs a little. She tells me she had been going through a very difficult period in her life when she came across a group of women "who had something different, even in their suffering. And they started talking about the presence of God in their lives in such a way that made me listen. And then it was, suddenly, God just is there. In one moment, it was something impossible for me to think. In the other moment, it was there."

Conversion—I had forgotten about that. And yet it may be the key to understanding the power and potential of "Laudato sì." Pope Francis devotes an entire chapter of the encyclical to the need for an "ecological conversion" among Christians, "whereby the effects of their encounter with Jesus Christ become evident in their relationship with the world around them. Living our vocation to be protectors of God's handiwork is essential to a life of virtue; it is not an optional or a secondary aspect of our Christian experience."

An evangelism of ecology, I realize, is what I have been witnessing take shape during the past three days in Rome—in the talk of "spreading the good news of the encyclical," of "taking the Church on the road," of a "people's pilgrimage" for the planet, in Miranda laying out plans to spread the encyclical in Brazil through radio ads, online videos, and pamphlets for use in parish study groups.

A millennia-old engine designed to proselytize and convert non-Christians is now preparing to direct its missionary zeal inward, challenging and changing foundational beliefs about humanity's place in the world among the already faithful. In the closing session, Father McDonagh proposes "a three-year synod on the encyclical" to educate Church members about this new theology of interconnection and "integral ecology."

Many have puzzled over how "Laudato sì" can simultaneously be so sweepingly critical of the present and yet so hopeful about the future. The Church's faith in the power of ideas and its fearsome capacity to spread information globally go a long way toward explaining this tension. People of faith, particularly missionary faiths, believe deeply in something that a lot of secular people aren't so sure about: that all human beings are capable of profound change. They remain convinced that the right combination of argument, emotion, and experience can lead to life-altering transformations. That, after all, is the essence of conversion.

The most powerful example of this capacity for change may well be Pope Francis's Vatican. And it is a model not for the Church alone. Because if one of the oldest and most tradition-bound institutions in the world can change its teachings and practices as radically, and as rapidly, as Francis is attempting, then surely all kinds of newer and more elastic institutions can change as well.

And if that happens—if transformation is as contagious as it seems to be here—well, we might just stand a chance of tackling climate change.

POSTSCRIPT

Of all the pieces in this collection, I found rereading this one to be most troubling. Because for all Pope Francis's courage in calling out world governments on their ecological dereliction and their brutal disregard for migrant life, the Vatican has still failed to hold its own leaders accountable for the systematic sexual abuse of children and nuns and the deliberate cover-ups of those crimes. This denial of justice has tormented many of the Church's faithful and has undermined Francis's moral authority to lead on other issues, including the climate crisis. If nothing else, this should be a reminder of the urgency of an intersectional approach to social and political change: if we pick and choose which urgent crises to take seriously, the end result will be an inability to effect change on any of them. Only a fearless and holistic approach, which sacrifices no issue on the altar of any other, will deliver the deep transformation we need.

LET THEM DROWN: THE VIOLENCE OF OTHERING IN A WARMING WORLD

A culture that places so little value on black and brown lives that it is willing to let human beings disappear beneath the waves, or set themselves on fire in detention centers, will also be willing to let the countries where black and brown people live disappear beneath the waves, or desiccate in the arid heat.

MAY 2016
EDWARD W. SAID LONDON LECTURE

EDWARD SAID WAS NO TREE HUGGER. DESCENDED FROM TRADERS, ARTISANS, and professionals, the great anticolonial intellectual once described himself as "an extreme case of an urban Palestinian whose relationship to the land is basically metaphorical." In *After the Last Sky*, his meditation on the photographs of Jean Mohr, he explored the most intimate aspects of Palestinian lives, from hospitality to sports to home décor. The tiniest detail (the placing of a picture frame, the defiant posture of a child) provoked a torrent of insight from Said. Yet, when he was confronted with images of Palestinian farmers (tending their flocks, working the fields), the specificity suddenly evaporated. Which crops were being cultivated? What was the state of the soil? The availability of water? Nothing was forthcoming. "I continue to perceive a population of poor, suffering, occasionally colorful peasants, unchanging and collective," Said confessed. This perception was "mythic," he acknowledged—yet it remained.

If farming was another world for Said, those who devoted their lives to matters like air and water pollution appear to have inhabited another planet. Speaking to his colleague Rob Nixon, then at Columbia University, he once described environmentalism as "the indulgence of spoiled tree-huggers who lack a proper cause." But the environmental challenges of the Middle East are impossible to ignore for anyone immersed, as Said was, in its geopolitics. This is a region intensely vulnerable to heat and water stress, to sea-level rise and to desertification. A recent paper in *Nature Climate Change* predicts that unless we radically lower emissions and lower them fast, large parts of the Middle East will likely "experience temperature levels that are intolerable to humans" by the end of this century. And that's about as blunt as climate scientists get. Yet environmental issues in the region still tend to be treated as afterthoughts, or luxury causes. The reason is not ignorance, or indifference. It's just bandwidth. Climate change is a grave threat, but the most frightening impacts are a few years away. In the here and now, there are always far more pressing threats to contend with: military occupation, air assault, systemic discrimination, embargo. Nothing can compete with that; nor should it attempt to try.

There are other reasons that environmentalism might have looked like a bourgeois playground to Said. The Israeli state has long coated its nation-building project in a green veneer—it was a key part of the Zionist "back to the land" pioneer ethos. And in this context, trees, specifically, have been among the most potent weapons of land grabbing and occupation. It's not only the countless olive and pistachio trees that have been uprooted to make way for settlements and Israeli-only roads. It's also the sprawling pine and eucalyptus forests that have been planted over those orchards, and over Palestinian villages. The most notorious actor on this has been the Jewish National Fund, which, under its slogan, "Turning the Desert Green," boasts of having planted 250 million trees in Israel since 1901, many of them nonnative to the region. It has also directly funded key infrastructure for the Israeli military, including in the Negev Desert. In publicity materials, the JNF bills itself as just another green NGO, concerned with forest and water management, parks and recreation. It also happens to be the largest private landowner in the state of Israel, and despite a number of complicated legal challenges, it still refuses to lease or sell land to non-Jews.

I grew up in a Jewish community where every occasion (births and deaths, Mother's Day, bar mitzvahs) was marked with the proud purchase of a JNF tree in the name of the honored person. It wasn't until adulthood that I began to understand that those feel-good faraway conifers, certificates for which papered the walls of my Montreal elementary school, were not benign—not just something to plant and later hug. In fact, these trees are among the most glaring symbols of Israel's system of official discrimination, the one that must be dismantled if peaceful coexistence is to become possible.

The JNF is an extreme and recent example of what some call "green colonialism." But the phenomenon is hardly new; nor is it unique to Israel. There is a long and painful history in the Americas of beautiful pieces of wilderness being turned into conservation parks, and then that designation being used to prevent Indigenous people from accessing their ancestral territories to hunt and fish or simply to live. It has happened again and again. A contemporary version of this phenomenon is the carbon offset. Indigenous people from Brazil to Uganda are finding that some of the most aggressive land grabbing is being done by conservation organizations. A forest is suddenly rebranded a carbon offset and is put off-limits to its traditional inhabitants. As a result, the carbon offset market has created a whole new class of green human rights abuses, with farmers and Indigenous people being physically attacked by park rangers or private security when they try to access these lands. Said's comment about tree huggers should be seen in this context.*

And there is more. In the last year of Said's life, Israel's so-called separation barrier was going up, seizing huge swaths of the West Bank, cutting Palestinian workers off from their jobs, farmers from their fields, patients from hospitals—and brutally dividing families. There was no shortage of reasons to oppose the wall on human rights grounds. Yet, at the time, some of the loudest dissenting voices among Israeli Jews were not focused on any of that. Yehudit Naot, Israel's then-environment minister, was more worried about a report informing her that "The separation fence . . . is harmful to the landscape, the flora and fauna, the ecological corridors and the drainage of the creeks."

"I certainly don't want to stop or delay the building of the fence," she said, but "I am disturbed by the environmental damage involved." As the Palestinian activist Omar Barghouti later observed, Naot's "ministry and the National Parks Protection

*This context must be front of mind in the design and rollout of any contemporary Green New Deal. In order to avoid a replication of these colonial patterns, Indigenous knowledge and leadership will have to be embedded from the start, particularly when it comes to the ambitious tree-planting and ecological restoration projects that are badly needed to draw down carbon and provide storm protection on a large scale.

Authority mounted diligent rescue efforts to save an affected reserve of irises by moving it to an alternative reserve. They've also created tiny passages [through the wall] for animals."

Perhaps this puts the cynicism about the green movement in context. People do tend to be put off when their lives are treated with less respect than flowers and reptiles. And yet there is so much of Said's intellectual legacy that both illuminates and clarifies the underlying causes of the global ecological crisis, so much that points to ways we might respond that are far more inclusive than current campaign models: ways that don't ask suffering people to shelve their concerns about war, poverty, and systemic racism and first "save the world," but that instead demonstrate how all these crises are interconnected, and how the solutions could be, too. In short, Said may have had no time for tree huggers, but tree huggers must urgently make time for Said, and for a great many other anti-imperialist, postcolonial thinkers, because without that knowledge, there is no way to understand how we ended up in this dangerous place, or to grasp the transformations required to get us somewhere safer. So, what follows are some thoughts, by no means complete, about what we can learn from reading Said in a warming world.

He was and remains among our most achingly eloquent theorists of exile and homesickness, but Said's homesickness, he always made clear, was for a home that had been so radically altered that it no longer really existed. His position was complex: he fiercely defended the right of Palestinians to return, but never claimed that home was fixed. What mattered was the principle of respect for all human rights equally and the need for restorative justice to inform our actions and policies. This perspective is deeply relevant in our time of eroding coastlines, of nations disappearing beneath rising seas, of the coral reefs that sustain entire cultures being bleached white, of a balmy Arctic. This is because the state of longing for a radically altered homeland, a home that may not even exist any longer, is something that is being rapidly, and tragically, globalized.

In March 2016, two major peer-reviewed studies warned that sea level rise could happen significantly faster than previously believed. One of the authors of the first study was James Hansen, perhaps the most respected climate scientist in the world. He warned that, on our current emissions trajectory, we face the "loss of all coastal cities, most of the world's large cities and all their history"—and not in thousands of years from now but as soon as this century. In other words, if we don't demand radical change, we are headed for a whole world of people searching for a home that no longer exists.

Said helps us imagine what that might look like as well. He often invoked the Arabic word *sumud* ("to stay put, to hold on"), that steadfast refusal to leave one's land despite the most desperate eviction attempts and even when surrounded by continuous danger. It's a word most associated with places like Hebron and Gaza, but it could be applied equally today to thousands of residents of coastal Louisiana who have raised their homes up on stilts so that they don't have to evacuate, or to Pacific Islanders whose slogan is "We are not drowning. We are fighting." In low-lying nations like the Marshall Islands and Fiji and Tuvalu, they know that so much sea level rise is already locked in from polar ice melt that their countries likely have no future. But they refuse to concern themselves with only the logistics of relocation, and wouldn't relocate even if there were safer countries willing to open their borders—a very big if, given that climate refugees aren't currently recognized under international law. Instead, they are actively resisting: blockading

Australian coal ships with traditional outrigger canoes, disrupting international climate negotiations with their inconvenient presence, demanding far more aggressive climate action. If there is anything worth celebrating in the Paris Climate Agreement—and sadly, there isn't enough—it has come about because of this kind of principled action: climate *sumud*.

But this only scratches the surface of what we can learn from reading Said in a warming world. He was, of course, a giant in the study of "othering," what is described in his 1978 book *Orientalism* as "disregarding, essentializing, denuding the humanity of another culture, people or geographical region." And once the other has been firmly established, the ground is softened for any transgression: violent expulsion, land theft, occupation, invasion. Because the whole point of othering is that the other doesn't have the same rights, the same humanity, as those making the distinction.

What does this have to do with climate change? Perhaps everything.

We have dangerously warmed our world already, and our governments still refuse to take the actions necessary to halt the trend. There was a time when many had the right to claim ignorance. But for the past three decades, since the Intergovernmental Panel on Climate Change was created and climate negotiations began, this refusal to lower emissions has been accompanied with full awareness of the dangers. And this kind of recklessness would have been functionally impossible without institutional racism, even if only latent. It would have been impossible without Orientalism, without all the potent tools on offer that allow the powerful to discount the lives of the less powerful. These tools—of ranking the relative value of humans—are what allow the writing off of entire nations and ancient cultures. And they are what allowed for the digging up of all that carbon to begin with.

• • •

Fossil fuels aren't the sole driver of climate change—there is also industrial agriculture and deforestation—but they are the biggest. And the thing about fossil fuels is that they are so inherently dirty and toxic that they require sacrificial people and places: people whose lungs and bodies can be sacrificed to work in the coal mines, people whose lands and water can be sacrificed to open-pit mining and oil spills. As recently as the 1970s, scientists advising the US government openly referred to certain parts of the country being designated "national sacrifice areas." Think of the mountains of Appalachia, blasted off for coal mining because so-called mountaintop removal coal mining is cheaper than digging holes underground. There must be theories of othering to justify sacrificing an entire geography—theories about the people who lived there being so poor and backward that their lives and culture don't deserve protection. After all, if you're a "hillbilly," who cares about your hills?

Turning all that coal into electricity required another layer of othering, too, this time for the urban neighborhoods next door to the power plants and refineries. In North America, these are overwhelmingly communities of color, black and Latino, forced to carry the toxic burden of our collective addiction to fossil fuels, with markedly higher rates of respiratory illnesses and cancers. It was in fights against this kind of environmental racism that the climate justice movement was born.

Fossil fuel sacrifice zones dot the globe. Take the Niger Delta, poisoned with an *Exxon Valdez* worth of spilled oil every year, a process that Ken Saro-Wiwa, before he was murdered by his government, called "ecological genocide." The executions of community leaders, he said, were "all for Shell." In my country, Canada, the decision to dig up the Alberta tar sands, a particularly heavy form of oil, has required the shredding of treaties with First Nations, treaties signed with the British Crown that guaranteed Indigenous peoples the right to continue to hunt, fish, and live traditionally on their ancestral lands. It required it because these rights are meaningless when the land is desecrated, when the rivers are polluted and the moose and fish are riddled with tumors. And it gets worse: Fort McMurray, the town at the center of the tar sands boom, where many of the workers live and where much of the money is spent, was just decimated by an infernal blaze, entire neighborhoods burned to the ground. It's that hot and that dry. And that excess heat has something to do with the interred substance being mined there.

Even without such dramatic events, this kind of resource extraction is a form of violence because it does so much damage to the land and water that it brings about the end of a way of life, a slow death of cultures that are inseparable from the land. Severing Indigenous people's connection to their culture used to be state policy in Canada, imposed through the forcible removal of Indigenous children from their families to boarding schools where their language and cultural practices were banned and where physical and sexual abuse was rampant. A recent Truth and Reconciliation Commission report on these residential schools called them part of a system of "cultural genocide."

The trauma associated with these layers of forced separation—from land, from culture, from family—is directly linked to the epidemic of despair ravaging so many First Nations communities today. On a single Saturday night in April 2016, in the community of Attawapiskat (population two thousand), *eleven people* tried to take their own lives. Meanwhile, DeBeers runs a diamond mine on the community's traditional territory; like all extractive projects, it had promised hope and opportunity.

"Why don't the people just leave?" the politicians and pundits ask. Many do. And that departure is linked, in part, to the thousands of Indigenous women in Canada who have been murdered or gone missing, often in big cities. Press reports rarely draw a connection between violence against women and violence against the land (often to extract fossil fuels), but one exists.

Every new government comes to power promising a new era of respect for Indigenous rights. They don't deliver because Indigenous rights, as defined by the UN Declaration on the Rights of Indigenous Peoples, include the right to refuse extractive projects, even when those projects fuel national economic growth. And that's a problem because growth is our religion, our way of life. So, even Justin Trudeau, Canada's woke young prime minister, is bound and determined to build new fossil fuel projects—new mines, new pipelines, and new export terminals—against the express wishes of Indigenous communities who don't want to risk their water, or participate in further destabilizing the climate.

The point is this: our fossil fuel–powered economy requires sacrifice zones. It always has. And you can't have a system built on sacrificial places and sacrificial people unless intellectual theories that justify their sacrifice exist and persist: from the Doctrine of Christian Discovery to Manifest Destiny to *terra nullius* to Orientalism, from backward hillbillies to backward Indians. We often hear climate change blamed on "human nature," on the inherent greed and shortsightedness of our species. Or we are told we have altered the earth so much and on such a planetary scale that we are now living in the Anthropocene, the age of man. These ways of explaining our current circumstances have a very specific, if unspoken meaning: that humans are a single type, that human nature can be essentialized to the traits that created this crisis. In this way, the systems that certain humans created, and other humans powerfully resisted, are completely let off the hook. Capitalism, colonialism, patriarchy—those sorts of systems.

Diagnoses like this also erase the very existence of human systems that organized life differently, systems that insist that humans must think seven generations in the future; must be not only good citizens but also good ancestors; must take no more than they need and give back to the land in order to protect and augment the cycles of regeneration. These systems existed and persist, against all odds, but they are erased every time we say that climate disruption is a crisis of "human nature" and that we are living in the "age of man."* And they come under very real attack when megaprojects are built, like the Gualcarque River hydroelectric dam in Honduras, a project that, among other things, stole the life of the land defender Berta Cáceres, who was assassinated in March 2016.

Some people insist that it doesn't have to be this bad. We can clean up resource extraction; we don't need to do it the way it's been done in Honduras and the Niger Delta and the Alberta tar sands.

*The idea that climate breakdown is not the doing of humanity as a homogenous unit but rather of specific imperial projects received strong historical reinforcement in early 2019. A team of scientists from University College London published a paper in Quaternary Science Reviews that made a persuasive case that the period of global cooling known as the "Little Ice Age," which took place in the 15-1600s, was partly caused by the genocide of Indigenous peoples in the Americas following European contact. The scientists argue that, with millions dead from disease and slaughter, huge swaths of land that had previously been used for agriculture were reclaimed by wild plants and trees, sequestering carbon and cooling the entire planet. "The Great Dying of the Indigenous Peoples of the Americas led to the abandonment of enough cleared land that the resulting terrestrial carbon uptake had a detectable impact on both atmospheric CO_2 and global surface air temperatures," the paper states. Prof. Mark Maslin, one of the coauthors, refers to this chillingly as a "genocide-generated drop in CO_2."

Except that we are running out of cheap and easy ways to get at fossil fuels, which is why we have seen the rise of fracking, deepwater drilling, and tar sands extraction in the first place. This, in turn, is starting to challenge the original Faustian pact of the industrial age: that the heaviest risks would be outsourced, offloaded, onto the other, the periphery abroad and inside our own nations. It's a pact that is becoming less and less tenable. Fracking is threatening some of the most picturesque parts of Britain as the sacrifice zone expands, swallowing up all kinds of places that imagined themselves safe. So, this isn't just about gasping at how ugly the vast tailing ponds are in Alberta. It's about acknowledging that there is no clean, safe, nontoxic way to run an economy powered by fossil fuels. There never was.

There is an avalanche of evidence that there is no peaceful way, either. The trouble is structural. Fossil fuels, unlike renewable forms of energy such as wind and solar, are not widely distributed but are highly concentrated in very specific locations, and those locations have a bad habit of being in other people's countries. Particularly that most potent and precious of fossil fuels: oil. This is why the project of Orientalism, of othering Arab and Muslim people, has been the silent partner of our oil dependence from the start—and inextricable, therefore, from the blowback from fossil fuel dependence that is climate change.

If nations and peoples are regarded as other—exotic, primitive, bloodthirsty, as Said documented in the 1970s—it is far easier to wage wars and stage coups when they get the crazy idea that they should control their own oil in their own interests. In 1953 it was the British-US collaboration to overthrow the democratically elected government of Mohammad Mosaddegh after he nationalized the Anglo-Persian Oil Company (now BP). In 2003, exactly fifty years later, it was another UK-US coproduction: the illegal invasion and occupation of Iraq. The reverberations from both interventions continue to roil our world, as do the reverberations from the successful burning of all that oil. The Middle East is now squeezed in the pincer of violence triggered by the quest for fossil fuels, on the one hand, and the impact of burning those fossil fuels on the other.

In his book, *The Conflict Shoreline*, the Israeli architect Eyal Weizman has a groundbreaking take on how these forces are intersecting. The main way we've understood the border of the desert in the Middle East and North Africa, he explains, is the so-called aridity line, areas where there is on average 7.8 inches (200 millimeters) of rainfall a year, which has been considered the minimum for growing cereal crops on a large scale without irrigation. These meteorological boundaries aren't fixed: they have fluctuated for various reasons, whether it was Israel's attempts to "green the desert" pushing them in one direction or cyclical drought expanding the desert in the other. And now, with climate change, intensifying drought can have all kinds of impacts along this line.

Weizman points out that the Syrian border city of Daraa falls directly on the aridity line. Daraa is where Syria's deepest drought on record brought huge numbers of displaced farmers in the years leading up to the outbreak of Syria's civil war, and it's where the Syrian uprising broke out in 2011. Drought wasn't the only factor in bringing tensions to a head, of course. But the fact that 1.5 million people were internally displaced in Syria as a result of the drought clearly played a role.

The connection between water and heat stress and conflict is a recurring, intensifying pattern that spans the aridity line: all along it you see places marked by drought, water scarcity, scorching temperatures, and military conflict—from Libya to Palestine to some of the bloodiest battlefields in Afghanistan, Pakistan, and Yemen.

And that's not all.

Weizman also discovered what he calls an "astounding coincidence." When you map the targets of Western drone strikes onto the region, you see that "many of these attacks—from South Waziristan through northern Yemen, Somalia, Mali, Iraq, Gaza and Libya—are directly on or close to the 200 mm aridity line."

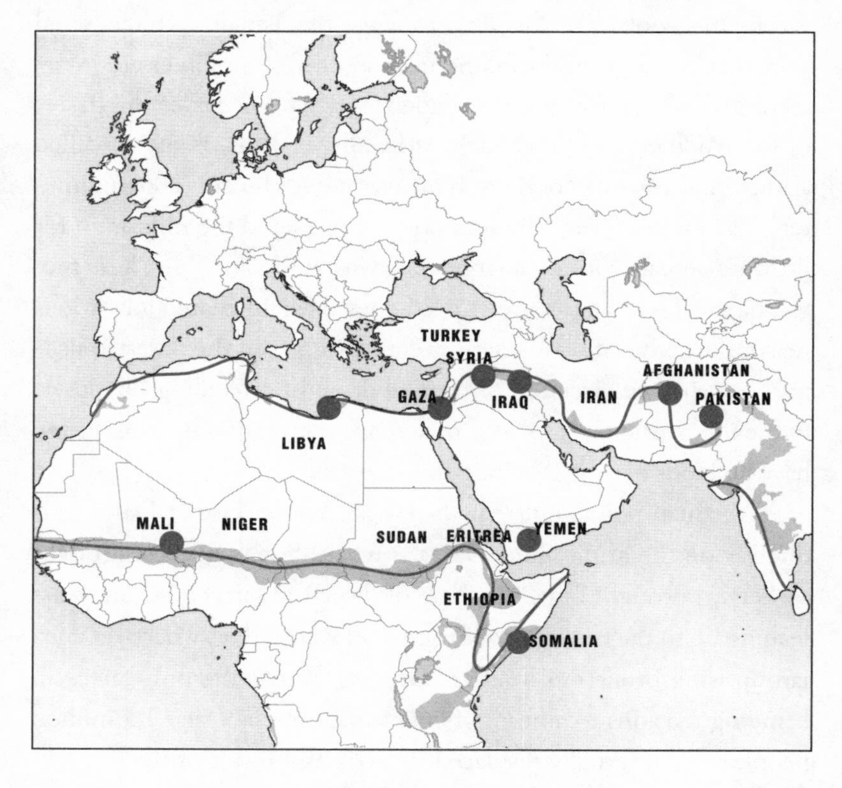

The red line on the map shows the aridity line; the red dots on the map represent some of the areas where strikes have been concentrated. To me, this is the most clarifying attempt yet to visualize the brutal landscape of the climate crisis.

All this was foreshadowed a decade ago in a US military report published by the Center for Naval Analyses. "The Middle East," it observed, "has always been associated with two natural resources, oil (because of its abundance) and water (because of its scarcity)." True enough. And now certain patterns have become quite clear: first, Western fighter jets followed that abundance of oil; now Western drones are closely shadowing the lack of water, as drought exacerbates conflict.

Just as bombs follow oil, and drones follow drought, so boats follow both: boats filled with refugees fleeing homes on the aridity line ravaged by war and drought. And the same capacity for dehumanizing the other that justified the bombs and drones is now being trained on these migrants, casting their need for security as a threat to ours, their desperate flight as some sort of invading army. Tactics refined on the West Bank and in other occupation zones are now making their way to North America and Europe. In selling his wall on the border with Mexico, Donald Trump likes to say, "Ask Israel, the wall works." Camps filled with migrants are bulldozed in Calais, France. Thousands of people drown in the Mediterranean every year.* And the Australian government detains survivors of wars and despotic regimes in camps on the remote islands of Nauru and Manus. Conditions are so desperate on Nauru that last month an Iranian migrant died after setting himself on fire to try to draw the world's attention. Another migrant, a twenty-one-year-old woman from Somalia, set herself on fire a few days later.

Malcolm Turnbull, the prime minister, warns that Australians

*In 2016, the year of this lecture, a record 5,143 migrants died in the crossing, according to the International Organization for Migration.

"cannot be misty-eyed about this" and "have to be very clear and determined in our national purpose." It's worth bearing Nauru in mind the next time a columnist in a Murdoch paper declares, as far-right commentator Katie Hopkins did last year, that it's time for Britain "to get Australian. Bring on the gunships, force migrants back to their shores and burn the boats."*

In another bit of symbolism, Nauru is one of the Pacific Islands very vulnerable to sea level rise. Its residents, after seeing their homes turned into prisons for others, will very possibly have to migrate themselves. Tomorrow's climate refugees have been recruited into service as today's prison guards.

We need to understand that what is happening on Nauru, and what is happening to it, are expressions of the same logic. A culture that places so little value on black and brown lives that it is willing to let human beings disappear beneath the waves, or set themselves on fire in detention centers, will also be willing to let the countries where black and brown people live disappear beneath the waves, or

*In recent years, Europe has adopted the Australian model with gusto. In an effort to restrict immigration, the Italian government has lavished the notoriously lawless Libyan Coast Guard with funding, training, logistical support, and equipment—all so that it is able to intercept the migrant boats before they reach European waters. Under this new system, the migrants who live—thousands still drown—are taken by force back to Libya and to what are frequently described as "concentration camps," places where torture, rape, and other forms of abuse are widespread. Meanwhile, international humanitarian organizations such as Médecins sans Frontières (MSF), which had previously saved thousands of migrants at sea, are facing criminalization and the seizure of their vessels. At the end of 2018, when MSF was forced to cease operations of its rescue ship *Aquarius*, MSF general director Nelke Manders remarked, "This is a dark day. Not only has Europe failed to provide search-and-rescue capacity, it has also actively sabotaged others' attempts to save lives. The end of *Aquarius* means more deaths at sea, and more needless deaths that will go unwitnessed."

desiccate in the arid heat. When that happens, theories of human hierarchy—that we must take care of our own first, that migrants are out to destroy "our way of life"—will be marshaled to rationalize these monstrous decisions. We are making this rationalization already, if only implicitly. Although climate change will ultimately be an existential threat to all of humanity, in the short term we know that it does discriminate, hitting the poor first and worst, whether they are abandoned on the rooftops of New Orleans during Hurricane Katrina or are among the thirty-six million who, according to the United Nations, are facing hunger due to drought in southern and East Africa.

This is an emergency, a present emergency, not a future one, but we aren't acting like it. The Paris Agreement commits to keeping warming below 2°C. It's a target that is beyond reckless. When it was unveiled in Copenhagen in 2009, many African delegates called it "a death sentence." The slogan of several low-lying island nations is "1.5 to Stay Alive." At the last minute, a clause was added to the Paris Agreement that says countries will pursue "efforts to limit the temperature increase to 1.5°C."

Not only is this nonbinding, but it is a lie: we are making no such efforts. The governments that made this promise are now pushing for more fracking and more mining of the highest-carbon fossil fuels on the planet, actions that are utterly incompatible with capping warming at 2°C, let alone 1.5°C. This is happening because the wealthiest people in the wealthiest countries in the world think they are going to be okay, that someone else is going to eat the biggest risks, that even when climate change turns up on their doorstep, they will be taken care of.

When they're proven wrong, things get even uglier. We had

a vivid glimpse into that future when the floodwaters rose in England in December 2015, inundating sixteen thousand homes. These communities weren't only dealing with the wettest December on record. They were also coping with the fact that the government has waged a relentless attack on the public agencies and the local councils that are on the front lines of flood defense. So, understandably, there were many who wanted to change the subject away from that failure. Why, they asked, is Britain spending so much money on refugees and foreign aid when it should be taking care of its own? "Never mind foreign aid," we read in the *Daily Mail*. "What about national aid?"

"Why," a *Telegraph* editorial demanded, "should British taxpayers continue to pay for flood defenses abroad when the money is needed here?" I don't know—maybe because Britain invented the coal-burning steam engine and has been burning fossil fuels on an industrial scale longer than any nation on earth? But I digress. The point is that this could have been a moment to understand that we are all affected by climate change and must take action together and in solidarity with one another. But it wasn't, because climate change isn't just about things getting hotter and wetter: under our current economic and political order, it's about things getting meaner and uglier.

The most important lesson to take from all this is that there is no way to confront the climate crisis as a technocratic problem, in isolation. It must be seen in the context of austerity and privatization, of colonialism and militarism, and of the various systems of othering needed to sustain them all. The connections and intersections between them are glaring, and yet so often, resistance to them is highly compartmentalized. The anti-austerity people rarely talk about climate change; the climate change people rarely talk about war or occupation. Too many of us fail to make the connection between the guns that take black lives on the streets of US cities and in police custody and the much larger forces that annihilate so many black lives on arid land and in precarious boats around the world.

Overcoming these disconnections, strengthening the threads tying together our various issues and movements, is, I would argue, the most pressing task of anyone concerned with social and economic justice. It is the only way to build a counterpower sufficiently robust to win against the forces protecting the highly profitable but increasingly untenable status quo. Climate change acts as an accelerant to many of our social ills (inequality, wars, racism, sexual violence), but it can also be an accelerant for the opposite, for the forces working for economic and social justice and against militarism. Indeed, the climate crisis, by presenting our species with an existential threat and putting us on a firm and unyielding science-based deadline, might just be the catalyst we need to knit together a great many powerful movements bound together by a belief in the inherent worth and value of all people and united by a rejection of the sacrifice zone mentality, whether it applies to peoples or to places.

We face so many overlapping and intersecting crises that we can't afford to fix them one at a time. We need *integrated* solutions, solutions that radically bring down emissions while creating huge numbers of good, unionized jobs and delivering meaningful justice to those who have been most abused and excluded under the current extractive economy.

Said died the year Iraq was invaded, living to see its libraries and museums looted—and its oil ministry faithfully guarded. Amid these outrages, he found hope in the global antiwar movement and in new forms of grassroots communication opened up by technology; he noted "the existence of alternative communities across the globe, informed by alternative news sources, and keenly aware of the environmental, human rights and libertarian impulses that bind us together in this tiny planet." Yes, "environmental"—his vision even had a place for tree huggers.

I was reminded of those words recently while I was reading up on England's floods. Amid all the scapegoating and finger-pointing, I came across a post by a man called Liam Cox. He was upset by the way some in the media were using the disaster to rev up anti-foreigner sentiment, and he said so:

> I live in Hebden Bridge, Yorkshire, one of the worst affected areas hit by the floods. It's shit, everything has gotten really wet. However . . . I'm alive. I'm safe. My family are safe. We don't live in fear. I'm free. There aren't bullets flying about. There aren't bombs going off. I'm not being forced to flee my home and I'm not being shunned by the richest country in the world or criticized by its residents.
>
> All you morons vomiting your xenophobia . . . about how money should only be spent "on our own" need to look at yourselves closely in the mirror. I request you ask yourselves a very important question . . . Am I a decent and honourable human being? Because home isn't just the UK, home is everywhere on this planet.

I think that makes for a very fine last word.

THE LEAP YEARS: ENDING THE STORY OF ENDLESSNESS

When you have gone as badly off course as we have, moderate actions don't lead to moderate outcomes. They lead to dangerously radical ones.

SEPTEMBER 2016
LAFONTAINE-BALDWIN LECTURE, TORONTO

MY DIRTY CANADIAN SECRET—AND PLEASE DON'T TURF ME OUT OF THIS LOVELY hall for it—is that I'm actually American. I even brought my passport to prove it. I also have a Canadian one. By law, when I travel into the United States, I have to show the one with the eagle on it. And when I travel back home to Toronto, I show the one with the elaborate coat of arms with lots of British things on it (as well as a smattering of maple leaves that you can't really make out).

Let me explain this duality. My parents are both Americans, born in the United States, which at the time gave their kids de facto US citizenship. I, on the other hand, was born in Montreal, and I have lived in Canada my entire life—except for a handful of years before I was five. In my twenties and thirties, I was always very clear that my Americanness was a technicality, not an identity.

I rarely mentioned it, even to good friends. I checked the "Canada" box on forms and stood in the "Canada" line at the airport. And when I gave speeches and interviews in the United States, I said "your government," *not* "our government." And even though my parents told me I was entitled to one, I never applied for a US passport. I sort of liked not having physical proof of my Americanness.

So, what changed? In 2011, I was in Washington, DC, at a protest against the Keystone XL Pipeline, which, if built, would carry tar sands bitumen from Alberta to the Gulf Coast.* The action in Washington included civil disobedience, a decision, by thousands of people over a two-week period, to peacefully trespass in front of the White House and get arrested. No non-Americans were supposed to participate in the civil disobedience part of the action, since getting arrested in the United States can have serious implications for your ability to reenter the country.

But something happened on that day in Washington: a delegation of Indigenous people from Northern Alberta, whose traditional territory has been badly damaged by oil and gas development, decided to risk the repercussions and get arrested anyway. Impulsively, and without warning my husband, Avi (which he always reminds me of), and I decided to join them.

It was a good day. I met some amazing people in the paddy wagon and at the bar afterward. After we were all released, it occurred to me that I might have trouble getting that US passport now. I was fine with it but decided to see what would happen if I tried. Much to my surprise, it worked, and that's how I finally got a US passport in my forties.

*Despite Donald Trump's multiple attempts to push through the $8 billion pipeline via executive order, it remained tied up in court challenges as this book went to press.

So, that explains the American part, but it doesn't explain why my American family came to Canada in the first place. That's a whole other story, also involving jail. It was 1967, my father was finishing up medical school, and both my parents were active against the war in Vietnam. Like many of his peers, my father did everything he could to avoid the draft: he filed for Conscientious Objector status, he tried to find an alternative form of service, you name it. It didn't work, and he found himself faced with a choice between going to Vietnam, going to jail, or going to Canada. So . . . here we are.

On car trips, my parents would regale us kids with stories of their escape, which to us sounded like a high-octane thriller: the letter from the army, the shotgun wedding, the secrecy to keep others from being implicated in their crime. We heard about how they boarded a late-night flight that landed in Montreal at midnight because they had heard that the Francophone anti-American customs agents worked the graveyard shift. Then—phew—they were waved through. Here's how my father recalls their arrival: "In twenty minutes we were landed immigrants, on the path to Canadian citizenship!"

Growing up in Canada with American lefty parents gave me a pretty rosy picture of this country. I heard a lot about the reasons they'd left the United States: the militarism, the jingoism, the millions without health insurance. And a lot about the things that drew them to, and kept us in, Canada. Like Prime Minister Pierre Trudeau declaring Canada "a refuge from militarism," universal public health care, public support for media and the arts. (My mother landed a staff job at the National Film Board where she was paid by the government to make subversive feminist documentaries.) In retrospect, it was a tiny bit like growing up in one of those Michael Moore films that show Canada as a utopian, alter-USA, where no one locks their doors, and no one gets shot, and no one waits to see a doctor, and everyone is super nice to each other *all the time*.

It wasn't quite that cartoonish. But there was a lot of stuff missing in the American-filtered stories of Canada that shaped my childhood and my own national pride. I now know, for instance, that while Canadians were feeling righteous about not joining the war in Vietnam and welcoming draft dodgers, Canadian companies were selling weapons and billions in other materials to supply the US war effort, including napalm and Agent Orange. Having it both ways is something of a Canadian military tradition. We did it again in 2003, when Canada very publicly did not participate in the 2003 invasion of Iraq because the attack did not have UN approval—and then, far less publicly, supported the subsequent occupation with exchange officers and warships.

It can be painful to look too closely at the stories that make us feel good, especially when they are part of the intimate narratives that mold our identities. I struggle with this still. I agree with my parents that our health care system and support for public media and the arts are part of what make us different from the United States. But it's also true that these institutions and traditions are deeply diminished after decades of neglect. These days, my father spends much of his retirement working to defend our public health system against encroaching US-style privatizations.

There is something else about my happy Canadian story that needs some poking. That frictionless experience at the airport—twenty minutes to landed immigrant status. That very likely had a lot to do with the fact that my parents, like many of the draft dodgers, were white, middle class, and college educated. These were not the only people fleeing war that Canada welcomed in this period; we also received sixty thousand Vietnamese refugees.

But this window of openness was relatively brief and a response, in part, to our shameful refusal to accept Jewish refugees during the Second World War. In recent decades, black and brown people on the bombardment end of illegal wars, including the wars we have helped fuel with weapons or soldiers or both, most certainly do not get landed immigrant status in twenty minutes, free to start work on Monday morning. Thousands are thrown in jail for years, charged with absolutely no crime. Many are in maximum-security prisons, with no idea when they will be released, a practice that has been repeatedly criticized by the United Nations.

The stories we tell about who we are as a nation, and the values that define us, are not fixed. They change as facts change. They change as the balance of power in society changes. Which is why regular people, not just governments, need to be active participants in this process of retelling and reimagining our collective stories, symbols, and histories.

And this is happening, too. For instance, all around Toronto, where we gather, the Ogimaa Mikana Project has been replacing official street signs with their Anishinaabe-language versions. They also put up a billboard near where I live reminding passersby that our rapidly gentrifying neighborhood is the subject of the Dish with One Spoon Wampum Belt Covenant, an agreement to peaceably share and care for the land and water. It's a very public attempt to change the collective story or, more accurately, to lift up older stories that are still alive but are usually drowned out by the barrage of louder, newer messages we receive day in and day out.

Interrogating the stories we have long taken for granted is healthy, especially the comforting ones. When the narratives and mythologies still feel helpful and true, resolving to do more to live up to them is also healthy. But when they no longer serve us, when they stand in the way of where we need to go, then we need to be willing to let them rest and tell some different stories.

THE LEAP

With that in mind, I want to share with you some reflections on one attempt at collective retelling—and how it clashed with some very powerful national narratives at the heart of the global ecological crisis. It's a project that I have been involved with called the Leap Manifesto. Many of you are aware of it. I know some of you have signed it. But the story behind the Leap is not very well known.*

The Leap came out of a meeting that was held in Toronto in May of 2015, attended by sixty organizers and theorists, from across the country, representing a cross section of movements: labor, climate, faith, Indigenous, migrant, women, antipoverty, anti-incarceration, food justice, housing rights, transit, and green tech. The catalyst for the gathering was a sudden drop in the price of oil, which had sent shock waves through our economy because of its reliance on revenues from the export of high-priced oil. The focus of our meeting was how we could harness that economic shock, which vividly showed the danger of hanging your fortunes on volatile raw resources, to kick-start a rapid shift to a renewables-based economy. For a long time, we had been told that we had to choose between a healthy environment and a strong economy; when the

*The Leap was, in many ways, a kind of proto-Green New Deal plan, an attempt to link ambitious climate action with a transition to a much fairer and more inclusive economy. The strengths and weaknesses of our experiment may be useful as the Green New Deal model is attempted in different countries.

price of oil collapsed, we ended up with neither. It seemed like a good moment to propose a radically different model.

At the time that we met, a federal election campaign was just gearing up, and it was already clear that none of the major parties was going to run on a platform of a rapid shift to a post-carbon economy. Both the Liberals and the New Democratic Party (NDP), then vying to unseat the governing Conservatives, were following the playbook that you needed to signal your "seriousness" and pragmatism by picking at least one major new oil pipeline and cheering for it. There were vague promises being offered on climate action but nothing guided by science, and nothing that presented a transition to a green economy as a chance to create hundreds of thousands of good jobs for the people who need them most.

So, we decided to intervene in the debate and write a kind of people's platform, the sort of thing we wished we could vote for but that wasn't yet on offer. And as we sat in a circle for two days and looked each other in the eye, we realized that this was new territory for contemporary social movements. We had all, or most of us, been part of broad coalitions before, opposing a particularly unpopular politician's austerity agenda, or coming together to fight against an unwanted trade deal or an illegal war.

But those were "no" coalitions, and we wanted to try something different: a "yes" coalition. And that meant we needed to create a space to do something we never do, which is dream together about the world that we actually want.

I am sometimes described as the author of the Leap Manifesto, but that's not true. My role was to listen, and notice the common themes. One of the clearest themes was the need to move from the national narrative that many of us had grown up with, that was based on a supposedly divine right to endlessly extract from the natural world as if there were no limit and no such thing as a breaking point. What we needed to do, it seemed to us, was set that story aside and tell a different one based on a duty to care: to care for the land, water, air—and to care for one another.

Largely because of the diversity in the room, we were also conscious that if we wanted a genuinely broad "yes" coalition, we couldn't fall back on a vision that was nostalgic or backward looking—a prelapsarian yearning for a seventies-era nation that never respected Indigenous sovereignty and that excluded the voices of so many communities of color, that often put too much faith in a centralized state and never actually reckoned with ecological limits.

So, rather than looking back, we started our platform with where we wanted to end up:

"We could live in a country powered entirely by renewable energy, woven together by accessible public transit, in which the jobs and opportunities of this transition are designed to systematically eliminate racial and gender inequality. Caring for one another and caring for the planet could be the economy's fastest growing sectors. Many more people could have higher wage jobs with fewer work hours, leaving us ample time to enjoy our loved ones and flourish in our communities."

The idea was to first paint a clear picture of where we wanted to go, and then get into the nitty-gritty of what it would take to get to that place. But before I get into those details, I want to return to the challenge of official stories.

You can tell from the name, the Leap, that it is about big and rapid change. That's why we chose it as our title: Because we know that when it comes to climate change, we have procrastinated for so long and made the problem so much worse that small steps, even if they are in the right direction, are still going to land us in a very deep hole. However, by framing our project as one of transformation, not incrementalism, we also put ourselves in a head-on collision with a story cherished by a lot of powerful interests in this country: that we are a moderate people, steady-as-she-goes kind of folks. In a world of hotheads, we like to tell ourselves that we split the difference, choose the middle path. No sudden movement for us, and certainly no leaping.

Now, it's a very nice story, and moderation is an asset in all sorts of circumstances. It's a good approach to alcohol consumption, for instance, and hot fudge sundaes. The problem, and the reason we chose this very un-moderate title quite consciously, is that when it comes to climate change, incrementalism and moderation are actually a huge problem. Because they will lead us, ironically, to a very extreme, hot, and cruel future. When you have gone as badly off course as we have, moderate actions don't lead to moderate outcomes. They lead to dangerously radical ones.

This was not always the case. The first intergovernmental meeting to talk about the climate crisis and the need for industrialized nations to lower emissions was held in 1988. Canada hosted it. It took place in this very city, and it came up with some fantastic recommendations. If we had listened to them, if we had all started cutting our emissions three decades ago, we could have taken it nice and slow: chipped away at our carbon footprint, knocked it down a couple of percentage points a year. A very moderate, gradual, centrist type of phaseout.

We didn't do that. We—not just our country but virtually every wealthy and fast-developing nation—did not do that. In fact, as governments met year after year to talk about lowering emissions, emissions went up by more than 40 percent. Here in Canada, we opened up huge new fossil fuel frontiers, and developed technology to dig up some of the highest-carbon oil on the planet. We didn't back off on the drivers of climate disruption; we doubled down. That was not very moderate—it was actually quite extreme.

So, now the problem is much worse. Worse because emissions have exploded, so we have to cut them far more deeply to bring them to safe levels. And worse because we have no time left, so we need to start these cuts immediately. That's what happens when you kick the can down the road enough times. You run out of road.

So, now we really do have to take radical action. Sudden and sweeping action, never mind how profoundly it conflicts with those comforting stories we tell ourselves about our centrist souls. Call it what you want: a Green New Deal, the Great Transition, a Marshall Plan for Planet Earth. But make no mistake: This is not an add-on, one more item on a governmental to-do list; nor is the planet some special interest to satisfy. The kind of transformation that is now required will happen only if it is treated as a civilizational *mission*, in our country and in every major economy on earth.

One thing we were very conscious of when we drafted the Leap Manifesto is that emergencies are vulnerable to abuses of power, and progressives are not immune to this by any means. There is a long and painful history of environmentalists, whether implicitly or explicitly, sending the message that "Our cause is so big, and so urgent, and since it encompasses everyone and everything, it should take precedence over everything and everyone else." Between the lines: "First we'll save the planet and then we will worry about poverty, police violence, gender discrimination, and racism."

In fact, that is a great way to build a very small, weak, and homogenous movement. Because poverty, war, racism, and sexual violence are *all* existential threats if you and your community are in the crosshairs. So, inspired by the climate justice movement growing around the world, we tried something else. We resolved that if we were going to radically change our economy, to make it a lot cleaner in the face of climate catastrophe, then we had to seize this opportunity to make it a lot fairer at the same time, on all these different fronts. That way nobody was being asked to choose between which existential threat mattered most to them. I'll give you a few quick examples.

Unsurprisingly, for a climate-focused document, we called for big investments in green infrastructure: renewables, efficiency, transit, high-speed rail. All of it to get to a 100 percent renewable economy by mid-century and 100 percent renewable energy well before that. We knew that all this would be a huge job creator—investing in these sectors creates six to eight times more jobs than putting that money in oil and gas. So, we called for public money to retrain those workers who face losing their jobs in extractive sectors, so that they are ready to work in the next economy, and the unions around the table told us that it was crucial for workers to be democratically involved in designing those retraining programs. So, that's all in the platform: basic principles of a justice-based transition.

But we also wanted something more. When we talk about "green jobs"—and we talk about them a lot—most of us picture a guy in a hard hat putting up a solar array. Sure, that is one kind of green job, and we need lots of them. But there are plenty of other jobs that are already low-carbon. For instance, looking after elderly and sick people doesn't burn a lot of carbon. Making art doesn't burn a lot of carbon. Teaching kids is low-carbon. Day care is low-carbon. And yet this work, overwhelmingly done by women, tends to be undervalued, underpaid, and is frequently the target of government cutbacks. So, we decided to deliberately extend the usual definition of a green job to anything useful and enriching to our communities that doesn't burn a lot of fossil fuels. As one participant said, "Nursing is renewable energy. Education is renewable energy." Moreover, this kind of work makes our communities stronger, more humane, and, therefore, better able to navigate the shocks that are headed our way in a climate-disrupted future.

Another key plank in the Leap Manifesto is what is known as "energy democracy," the idea that renewable energy, whenever possible, should be public- or community-owned and controlled so that the profits and benefits of new industries are far less concentrated than they are with fossil fuels. We were inspired by Germany's energy transition, which has seen hundreds of cities and towns taking back control over their energy grids from private companies, as well as an explosion of green energy cooperatives, where the profits from power generation stay in the community to pay for essential services.

But we decided that we need more than energy democracy, that we also need energy justice, even energy reparations. Because the way energy generation and other dirty industries have developed over the past couple of centuries has forced the poorest communities to bear a vastly disproportionate share of the environmental burdens while deriving far too little of the economic benefits. Which is why the Leap states that "Indigenous Peoples and others on the front lines of polluting industrial activity should be first to receive public support for their own clean energy projects."

Some find these kinds of connections daunting. Lowering emissions is hard enough, we are told—why weigh it down by trying to fix so much else at the same time? Our response is that if we are going to repair our relationship to the land by shifting away from endless resource extraction, why wouldn't we begin to repair our relationships with one another in the process? For a very long time, we have been offered policies that amputate the ecological crises from the economic and social systems that are driving them. That is precisely the model that has failed to yield results. Holistic transformation, on the other hand, has never been tried on a national scale.

Another example. The Leap explicitly acknowledges the role

that our government's foreign policies have played, and continue to play, in pushing people to leave their homes and seek asylum in other countries. Some are pushed by the dire economic impacts of trade deals that our government supported, some by mines that our companies have built. Some are pushed by wars that our government helped wage or fund.

All these—the trade deals, the wars, the mines—are major contributors to the increase in global greenhouse gas emissions, and now climate change itself is also forcing people to leave their homes. Which is why we decided to reframe migrant rights as a climate justice issue. We clearly stated that we need to open our borders to many more migrants and refugees, and that all workers, regardless of immigration status, should have full labor rights and protections. We need to do this not out of charity or as an expression of the goodness in our hearts, but because climate change, in its global complexity, teaches us that our fates are, and always have been, interconnected. Underneath it all, this is about what kind of people we want to be as the impacts of our collective action become undeniable. It's a moral and spiritual question as much as an economic and political one.

We knew that the greatest obstacle our platform would face was the force of austerity logic—the message we have all received, over decades, that governments are perpetually broke, so why even bother dreaming of a genuinely equitable society? With this in mind, we worked closely with a team of economists to come up with a parallel document that showed exactly how we would raise the revenues to pay for our plan.

Before releasing the platform to the public, we approached many organizations and high-profile individuals. Again and again, we heard: Yes. This is who we want to be. Let's push our politicians. Canadian caution be damned. National icons stood with us without hesitation: Neil Young. Leonard Cohen. The novelist Yann Martel wrote back that it should be "shouted from the rooftops." This was a rare document that could be signed by Greenpeace, the head of the Canadian Labour Congress, and Indigenous elders like the famed Haida spokesperson and master carver Gujaaw. More than two hundred organizations in all.

THE BACKLASH

Given this initial enthusiasm, we were frankly a little surprised by what happened when we launched the platform into the wider world. "Shit storm" would be an understatement.

First, our former prime minister Brian Mulroney came out of retirement to declare the Leap "a new philosophy of economic nihilism" that "must be resisted and defeated." Then, after the NDP voted to endorse its spirit and debate its specifics, the sitting premiers of three provinces, from three different political parties, came out to denounce it. "Hundreds of towns would be wiped off the map. *Tomorrow*. And turned into ghost towns," one said. "An existential threat," said another. And finally, from the (now former) NDP Premier of Alberta: "A betrayal."

Interestingly, none of this seems to have had much of an impact at the grassroots. People keep adding their names to the platform. They keep starting local Leap chapters. And a poll conducted at the peak of the backlash found that a majority of Green, NDP, and Liberal voters supported the core ideas in the Leap Manifesto. Even 20 percent of Conservatives. I think this reveals a pretty interesting divide: A whole lot of people of different political persuasions read the Leap and thought it sounded eminently sensible, inspiring even. But our elites across party lines agreed that it sounded like the end of the world.

So, what can we make of that chasm? It was really just one line in the Leap that caused most of the uproar, the one that said that we can't build any more fossil fuel "infrastructure that locks us into increased extraction decades into the future." The "no pipelines" line.

Let's unpack that a little. From a scientific perspective, it's not at all controversial. In Paris, governments negotiated a climate treaty that pledged to keep warming below 2°C while pursuing "efforts to limit the temperature increase to 1.5°C." (It was Justin Trudeau's team that fought to get that more ambitious language in there.)

To put that in perspective, we have already warmed the planet by roughly 1°C from where we were before humans starting burning coal on an industrial scale. So, if 1.5–2°C is our goal, then that puts us on a very constrained carbon budget. Staying within it—and scientists have been very clear on this—requires that we leave a whole lot of our current carbon reserves in the ground. For particularly dirty forms of fossil fuel, like Alberta's bitumen, it means about 85–90 percent of it has to stay in the ground. This is peer-reviewed research that has been published in the journal *Nature* and elsewhere; it's not contested.

Same goes for opening up new fossil fuel frontiers with technologies like fracking. And our politicians don't dispute it. They admit that their current emission-reduction targets—and this is true not just for Canada—take us way beyond the temperature goals they set in Paris. They do not add up to a carbon budget of 1.5–2°C. They add up to warming of 3–4°C—and that's if we manage to meet those targets. A big if.

We can have a debate about whether it is worth doing the very difficult things necessary to keep from warming the planet by 3–4°C (which, by the way, climate scientists have said is incompatible with anything you could describe as organized civilization). It would be an interesting debate to have. But that is not the debate we are having. Instead, when people argue for climate policies that are guided by science and by our own government's very publicly stated goals, they are basically told to shut up and stop destroying the country.

A UNIQUELY CONSTRICTED DEBATE

This is not true everywhere. Other countries are moving ahead with some of the policies that actually reflect scientific realities. Germany and France have both banned fracking, for instance. They both have a long way to go to bring their emissions in line with Paris Agreement temperature targets, but the aversion to talking about leaving carbon in the ground is not nearly as powerful in Europe as it is here. And we can't just blame this on the fact that we have a big oil and gas sector with lots of jobs on the line. Other countries do as well, and they are much farther along than us. Even the United Arab Emirates, a straight-up petrostate, is preparing for the end of oil, funneling tens of billions in oil wealth into new investments in renewables.

It's not just Canada that can't seem to have a rational debate about ecological limits. The debate is equally unhinged in Australia and the United States, with large segments of the political and pundit class denying the science outright—and the more this happens, the more the rest of the world is held back. I've been puzzling over what accounts for these geographic discrepancies. And I think it comes back to where we started: those official national narratives that tell countries what values define them, and the kind of power structures that these narratives nurture and maintain.

THE STORY OF ENDLESSNESS

When we launched the Leap, we hit up against a narrative that runs extremely deep, one that predates the founding of young countries like ours. It begins with the arrival of European explorers, at a time when their home nations had slammed into hard ecological limits: great forests gone, big game hunted to extinction.

It was in this context that the so-called New World was imagined as a sort of spare continent, to use for parts. (They didn't call it *New* France and *New* England by accident.)

And what parts! Here seemed to be a bottomless treasure trove—of fish, fowl, fur, giant trees, and, later, metals and fossil fuels. In North America and, later, in Australia, these riches covered territories so vast that it was impossible to fathom their boundaries. We were the place of endlessness—and whenever we began to run low, our governments just moved the frontier west.

The very existence of these lands appeared to come as a divine sign: Forget ecological boundaries. Thanks to this body-double continent, there seemed to be no way to exhaust nature's bounty. Looking back at early European accounts of what would become Canada, it becomes clear that explorers and early settlers truly believed that their scarcity fears were gone for good. The waters off the coast of Newfoundland were so full of fish that they "stayed the passage" of John Cabot's ships. For Quebec's Father Charlevoix in 1720, "The number of [cod] seems equal to that of the grains of sand that cover the bank." And then there were the great auks. The feathers of the penguin-like bird were coveted for mattresses, and on rocky islands, particularly off Newfoundland, they were found in huge numbers. As Jacques Cartier put it in 1534, there were islands "as full of birds as any field or meadow is of grasse."

Again and again, the words *inexhaustible* and *infinite* were used to describe the Eastern forests of great pines, the giant cedars of the Pacific Northwest, all manner of fish. Another common refrain is that the natural bounty is so great, there is really no point in worrying about managing this treasure trove to prevent depletion. There was so much that there was a glorious freedom to be careless. Thomas Huxley (the English biologist known as "Darwin's bulldog") told the 1883 International Fisheries Exhibition that "the cod fishery . . . are inexhaustible; that is to say nothing we do seriously affects the number of fish. Any attempt to regulate these fisheries seems consequently . . . to be useless."

That's a lot of famous last words, given what we now know. Given that by 1800 the great auks were completely wiped out. Given that beaver stocks began to crash in Eastern Canada soon after. Given that Newfoundland's supposedly inexhaustible cod was declared "commercially extinct" in 1992. As for our inexhaustible old-growth forests: virtually wiped out here in Southern Ontario. More than 91 percent of the biggest and best stands on Vancouver Island, gone.

Of course, a great deal of this is not unique to Canada. The early US economy was brutally extractive, too.* But there were

*As historian Greg Grandin recently argued in *The End of the Myth: From the Frontier to the Border Wall in the Mind of America*, the promise of advancing through an ever-expanding open frontier has been the primary way that US politicians have resolved social and ecological conflicts. Whenever the soil was depleted by careless farming, or one group of poor (white) immigrants demanded greater equality, the response was to violently seize yet more land from Native Americans and expand the sphere. But now the figurative wall has been reached, and there is no more frontier available, whether geographic, financial, or atmospheric. Grandin argues that Donald Trump and his border wall should be understood as a reaction to the crashing of the frontier myth: with no frontier left to conquer, Trump turns his full attention to hoarding US wealth for his chosen group, while locking out everyone else. This is why some key differences. The southern slave economy was based on the extraction of forced human labor, used to clear and cultivate land to feed the rapidly industrializing North. Though slavery did exist in Canada, our primary role in the transatlantic slave trade was as a supplier: Much of that supposedly endless cod was salted and shipped to the British West Indies (Jamaica, Barbados, British Guiana, Trinidad, Grenada, Dominica, Saint Vincent, and Saint Lucia). For wealthy plantation owners, cod was an invaluable source of cheap protein for enslaved Africans.

Our economic niche was always voraciously devouring wilderness—both animals and plants. Canada was an extractive company, the Hudson's Bay fur trading company, before it was a country. And that has shaped us in ways we have yet to begin to confront. But it does go some way toward explaining why it caused such an uproar when a group of us got together and said: Actually, we have hit the hard limits of what the earth can take; we have to leave resources in the ground, even when they are still profitable. The time for a new story, and a new economic model, is now.

Because such enormous fortunes have been built in North America purely on the extraction of wild animals, intact forest, interred metals, and fossil fuels, our economic elites have grown accustomed to seeing the natural world as their God-given larder. What we discovered with the Leap is that when someone or something (like climate science) comes along and challenges that claim, it doesn't feel like a difficult truth. It feels, as we learned, like an existential attack.

The economic historian Harold Innis (who never reckoned

---

outmoded national narratives cannot be left to die quietly. They need to be challenged with new stories that reflect how our knowledge has evolved and who we want to be—or else they'll turn septic and even more dangerous.

with Canada's crucial role in the slave trade) warned of this almost a century ago. Canada's extreme dependence on exporting raw natural resources, he argued, stunted our country's development at "the staples phase." This is true for large parts of the US economy as well—Louisiana and Texas for oil, West Virginia for coal. This reliance on raw resources makes economies intensely vulnerable to monopolies and to outside economic shocks. It's why the term *banana republic* is not considered a compliment.

Though Canada doesn't think of itself like that, and some regions have diversified, our economic history tells another story. Over the centuries, we have careened from bonanzas to busts. In the late 1800s, the beaver trade collapsed when European elites suddenly lost their taste for top hats made of pelts and moved on to smoother silk. Last year, the economy of Alberta went into free fall because of a sudden drop in the price of oil. We used to get yanked around by the whims of British aristocrats; now it's Saudi princes. I'm not sure that counts as progress.

The trouble isn't just the commodity roller coaster. It's that the stakes grow larger with each boom-bust cycle. The frenzy for cod crashed a species; the frenzy for tar sands oil and fracked gas is helping to crash the planet.

And yet despite these enormous stakes, we can't seem to stop. The dependence on commodities continues to shape the body politic of settler-colonial states like Canada, the United States, and Australia. And in all three countries, it will continue to confound attempts to heal relations with First Nations. That's because the basic power dynamic—our countries relying on the wealth embedded in their land—remains unchanged. For instance, when the fur trade was the backbone of wealth production in the northern parts of this continent, Indigenous culture and relationships to the land became a profound threat to the lust for extraction. (Never mind that there would have been no trade without Indigenous hunting and trapping skills.) Which is why attempts to sever those relationships to the land were so systematic. Residential schools were one part of that system. So were the missionaries who traveled with fur traders, preaching a religion that cast Indigenous cosmologies as sinful forms of animism—never mind, once again, that the worldviews they attempted to exterminate have a huge amount to teach us about how to regenerate the natural world, rather than endlessly deplete it.

Today in Canada, we have federal and provincial governments that talk a lot about "truth and reconciliation" for those crimes. But this will remain a cruel joke if nonindigenous Canadians do not confront the "why" behind those human rights abuses. And the why, as the official Truth and Reconciliation Commission report states, is simple enough: "The Canadian government pursued this policy of cultural genocide because it wished to divest itself of its legal and financial obligations to Aboriginal people and gain control over their land and resources."

The goal, in other words, was always to remove all barriers to unrestrained resource extraction. This is not ancient history. Across the country, Indigenous land rights remain the single greatest barrier to planet-destabilizing resource extraction, from pipelines to clear-cut logging. We're still trying to get the land, and what's underneath. We see it south of the border as well, in the Standing Rock Sioux's pitched struggle against the Dakota Access Pipeline. This was true two hundred years ago, and it is true today.

When governments talk of truth and reconciliation, and then push unwanted infrastructure projects, please remember this: There can be no truth unless we admit to the "why" behind centuries of abuse and land theft. And there can be no reconciliation when the crime is still in progress.

Only when we have the courage to tell the truth about our old stories will the new stories arrive to guide us. Stories that recognize that the natural world and all its inhabitants have limits. Stories that teach us how to care for each other and regenerate life within those limits. Stories that put an end to the myth of endlessness once and for all.

HOT TAKE ON A HOT PLANET

Ours is a culture of endless taking, as if there were no end and no consequences. A culture of grabbing and going. And now this grab-and-go culture has reached its logical conclusion. The most powerful nation on earth has elected a grabber in chief.

NOVEMBER 2016
SYDNEY PEACE PRIZE ACCEPTANCE SPEECH

AS I WAS MAKING NOTES FOR THIS LECTURE OVER THE PAST COUPLE OF WEEKS, I knew I really should be preparing two versions: the "Hillary wins" version and the "Trump wins" version.

Thing is, I couldn't quite bring myself to write the "Trump wins" version. My typing fingers went on strike. I knew I would be speaking to you a mere forty-eight hours after learning the US presidential election results, so in retrospect, I was grossly derelict in my duties. And I apologize if what follows seems rushed—it is rushed. A hot take, as they call it these days, on a hot planet.

If there is a single, overarching lesson in the Trump victory, perhaps it is this: Never, ever underestimate the power of hate. Never underestimate the power of direct appeals to power over "the other"—the migrant, the Muslim, black people, women.

Especially during times of economic hardship. Because when large numbers of white men find themselves frightened and insecure, and those men were raised in a social system built on elevating their humanity over all these others', a lot of them get mad. And there is nothing wrong in itself with being mad—there's lots to be mad about.

But within a culture that so systematically elevates some lives over others, anger makes many of those men, and women, putty in the hands of whatever demagogue of the moment is offering to deliver back an illusion of dominance, however fleeting. Build a wall. Lock 'em up. Deport them all. Grab 'em wherever you like and show 'em who's boss.

What other lessons can we take from our two-day-old reality that we now live in a world with a President Trump?

One lesson: that the economic pain is real and not going anywhere. Four decades of corporate neoliberal policies of privatization, deregulation, free trade, and austerity have made sure of that.

Another lesson: Leaders who represent that failed consensus are no match for the demagogues and neofascists who claim to be toppling it. They have nothing tangible to offer, and they are seen, quite correctly, as the people responsible for much of this economic dislocation.

Only a bold and genuinely redistributive agenda has a hope of speaking to that pain and directing it where it belongs: to the politician-purchasing elites who benefited so extravagantly from the auctioning off of public wealth; the polluting of land, air, and water; and the deregulation of the financial sphere.

But there is a deeper lesson that we must urgently learn from this week's events: If we want to win against the likes of Trump—and every country has its homegrown Trump—we must urgently confront and battle racism and misogyny, in our culture, in our movements, in ourselves. This cannot be an afterthought; it cannot be an add-on. It is central to how someone like Trump could rise to power. Many people said they voted for him *despite* his objectionable race and gender pronouncements. They liked what he had to say about trade and bringing back manufacturing and that he wasn't a "Washington insider."

Sorry, but that doesn't cut it. You cannot cast a ballot for someone who is so openly riling up race-, gender-, and physical ability–based hatreds unless, on some level, you think those issues aren't that important. You just can't do it. You can't do it unless you are willing to sacrifice "the other" for your (hoped-for) gain.

But this isn't just about Trump voters and the stories they may have told themselves. We have arrived at this dangerous moment also because of the stories about "the other" told on the progressive side of the political spectrum. Like the one that holds that when we fight against war and climate change and economic inequality, it will benefit black people and Indigenous people the most because they are most victimized by the current system.

That doesn't work, either. There is too long and too painful a track record of left movements for economic justice leaving workers of color, Indigenous people, and women's labor out in the cold.

To build a truly inclusive movement, there needs to be a truly inclusive vision that starts with and is led by the most brutalized and excluded. Rinaldo Walcott, a great Canadian writer and intellectual, issued a challenge a couple of months ago to white liberals and leftists. He wrote:

> Black people are dying in our cities, crossing oceans, in resource wars not of our making. . . . Indeed, it is obvious that Black peoples' lives are disposable in a way and fashion that is radically different from other groups globally.

> It is from this stark reality of marginalization that I want to propose that any new policy actions in the North American context ought to pass what I will call the Black test. The Black test is simple: it demands that any policy meet the requirement of ameliorating the dire conditions of Black peoples' lives. . . . When a policy does not meet this test, then it is a failed policy, from the first instance of its proposal.

That's worth thinking hard about. I know that my work has too often failed to pass that test. But now more than ever, those of us who talk about peace, justice, and equality must rise to that challenge.

When it comes to climate action, it's abundantly clear that we will not build the power necessary to win unless we embed justice—particularly racial but also gender and economic justice—at the center of our low-carbon policies. *Intersectionality*, the term coined by black feminist legal scholar Kimberlé Crenshaw, is the only path forward. We cannot play "my crisis is more urgent than your crisis"—war trumps climate; climate trumps class; class trumps gender; gender trumps race. That trumping game, my friends, is how you end up with a Trump.

Either we fight for a future in which everyone belongs, starting with those being most battered by injustice and exclusion today, or we will keep losing. And there is no time for that. Moreover, when we make these connections among issues (climate, capitalism, colonialism, white supremacy, and misogyny), there is a kind of relief. Because it actually *is* all connected, all part of the same story.

I was feeling this very intensely last week when I visited the Great Barrier Reef. I was there making a short film with the *Guardian* about this natural wonder, currently in the midst of a vast die-off, one directly linked to warming oceans.* Looking at a whole lot of bleached and dead coral, I found that most of my thoughts were about my four-year-old son, Toma, who still can't quite swim and will very likely never see a thriving reef in his lifetime.

There is no question that the strongest emotions I have about the climate crisis have to do with him and his generation—the tremendous intergenerational theft under way. I have flashes of sheer panic about the extreme weather we have already locked in for these kids. Even more intense than this fear is the sadness about what they won't ever know. They are growing up in a mass extinction, robbed of the cacophonous company of so many fast-disappearing life forms. It feels so desperately lonely.

But that wasn't all I was thinking about. Floating in the waters off Port Douglas, I also found myself thinking, as one does, about Captain James Cook. Thinking about all these forces that came together right around the time that the HMS *Endeavour* navigated those very waters.

As all you good students of Australian history know, Cook arrived in Queensland in 1770. Just six years later, the Watt commercial steam engine went on the market, a machine that massively accelerated the Industrial Revolution, now powered by a potent combination of slave labor in the colonies and coal fed into those

*In 2016 and 2017, triggered by warmer ocean temperatures, the Great Barrier Reef underwent a mass bleaching, which turned what had once been a riot of jewel-colored life into an eerie, ghostly white graveyard. Approximately half the vast reef's coral died in that period. In April 2019, new research was published revealing that the reef was not recovering. As *New Scientist* reported, "The amount of coral larvae on the reef in 2018 was down by 89 per cent on historical levels. 'Dead coral doesn't make babies,' says Terry Hughes of James Cook University in Australia, who led the work."

commercial steam engines. That same year, 1776, Adam Smith published *The Wealth of Nations*, the foundational text of contemporary capitalism—just in time for the United States to declare its independence from Britain.

Colonialism, slavery, coal, capitalism—all tightly bound up together in the span of six years, creating the modern world.

This country called Australia was born precisely at the dawn of fossil-fueled capitalism. We should connect the dots because they are connected—the land grabs, the fossil fuels that began changing our climate, the economic and social theories that rationalized it all. We are all living, in a very real sense, in Captain Cook's climate, or at least the one his fateful ocean voyages played an absolutely central part in creating.

One detail that particularly struck me in my research for this lecture: The HMS *Endeavour* didn't start life as a navy or scientific vessel, tasked with unlocking astrological and biological mysteries—and, in its spare time, claiming vast swaths of territory for the British Crown without Indigenous consent. No, the HMS *Endeavour* was built in 1764 to haul coal through British waterways. When the navy bought it, the boat had to be extensively (and expensively) retrofitted to be suited for Cook and Joseph Banks's voyage. It seems somehow fitting that the ship that laid claim to New South Wales and Queensland started life as a coal vessel.

Is it any wonder your government has an unnatural love affair with coal? Is it any wonder that not even the catastrophic bleaching of the Great Barrier Reef, one of the wonders of the world, has inspired Queensland's government to rethink its reliance on that black rock?

As Vandana Shiva said when accepting this prize six years ago, the roots of our crisis lie "in an economy which fails to respect ecological and ethical limits." Limits are a problem for our economic system. Ours is a culture of endless taking, as if there were no end and no consequences. A culture of grabbing and going.

And now this grab-and-go culture has reached its logical conclusion. The most powerful nation on earth has elected Donald Trump as its grabber in chief—a man who openly brags about grabbing women without their consent; who says about the invasion of Iraq, "We should have taken their oil," international law be damned.

This rampant grabbing is not just a Trump thing, of course. We have an epidemic of grabbing. Land grabbing. Resource grabbing. Even grabbing the sky by polluting so much that there is no atmospheric space left for the poor to develop.

And now we are hitting the wall of maximum grabbing. That's what climate change is telling us. That's what our endless wars are telling us. That's what Trump's electoral victory is telling us. That it's time to put everything we have into shifting from a culture of endless taking to a culture of consent and caretaking.

Caring for the planet, and for one another.

When I learned that I had been awarded the Sydney Peace Price for my climate work, I was incredibly honored. This is a prize that has gone to some of my personal heroes—Arundhati Roy, Noam Chomsky, Vandana Shiva, Desmond Tutu, among so many others. It's a very nice tribe to be a part of.

So, I was thrilled to receive the call. But after that wore off a bit, the doubts surfaced. One was: Why me? My writing builds on the work of so many thousands of climate justice activists around the world, many who have been at it for far longer than I. Another doubt was more practical: Can I really justify the transportation pollution required to accept an award for doing my bit to fight pollution? To be perfectly honest with you, I'm still not sure it can be justified.

But I consulted with Australian friends and colleagues. They pointed out that your government is the number-one coal exporter in the world, selling directly to those countries whose emissions are growing most rapidly. That you are well on your way to playing the same leading role for liquefied natural gas.

Even as other countries freeze and wind down their coal production, your prime minister is defiant. He says the plan is to stay the course with coal "for many, many decades to come"—long past the time when we all need to be off that dirty fuel if the Paris climate goals have a chance of being met. Earlier this week, I said that Australia stands increasingly alone in raising its sooty middle finger to the world. Unfortunately, I now have to amend that statement: Starting in January, when Donald Trump moves into the White House, Malcolm Turnbull will have some company. Ouch.

The Australian friends whom I consulted told me that having the megaphone that comes with this prize could help support their work—crucial efforts to stop new fossil fuel projects like the gargantuan Carmichael coal mine on Wangan and Jagalingou territory. And to stop the Northern Gas Pipeline, which would open up vast areas of the Northern Territory to industrial fracking.

This resistance is of global importance, because these mega projects concern massive pools of what we now call "unburnable carbon," carbon dioxide and methane that, if extracted and burned, will not only blow past Australia's paltry climate commitments but blow the global carbon budget as well. The math on this is very clear: In Paris, our governments (even yours) agreed to a goal of keeping warming below 2°C while pursuing "efforts to limit the temperature increase to 1.5°C."

That goal—and it's an ambitious one—places all of humanity

within the confines of a carbon budget. That's the total amount of carbon that can be emitted if we want to hit those targets and give island nations a fighting chance of surviving. And what we now know, thanks to breakthrough research from Oil Change International in Washington, DC, is that if we were to burn all the oil, gas, and coal from fields and mines already in production, we would very likely pass 2°C of warming and would certainly pass 1.5°C.

What we cannot do, under any circumstances, is precisely what the fossil fuel industry is determined to do and what your government is so intent on helping them do: dig *new* coal mines, open *new* fracking fields, and sink *new* offshore drilling rigs. All that needs to stay in the ground.

What we must instead do is clear: carefully wind down existing fossil fuel projects, at the same time as we rapidly ramp up renewables until we get global emissions down to zero globally by mid-century. The good news is that we can do it with existing technologies. The good news is that we can create millions of well-paying jobs around the world in the shift to a postcarbon economy—in renewables, in public transit, in efficiency, in retrofits, in cleaning up polluted land and water.

The better news is that as we transform how we generate energy, how we move ourselves around, how we grow our food and how we live in cities, we have a historic opportunity to build a society that is fairer on every front, and where everyone is valued. Here's how we do it. We make sure that, wherever possible, our renewable energy comes from community-controlled providers and cooperatives, so that decisions about land use are made democratically and profits from energy production are used to pay for much-needed services.

We know that our reliance on dirty energy over the past couple hundred years has taken its highest toll on the poorest and most vulnerable people, overwhelmingly people of color, many Indigenous. That's whose lands have been stolen and poisoned by mining. And it's poor urban communities who get the most polluting refineries and power plants in their neighborhoods.

So, we can and must insist that Indigenous and other frontline communities be first in line to receive public funds to own and control their own green energy projects—with the jobs, profits, and skills staying in those communities. This has been a central demand of the climate justice movement, led by communities of color. This is already starting to happen on an ad hoc basis. But too often, it is left to already underfunded communities to raise the finances. That is upside down: Climate justice means these communities are owed public funds as a drop in the ocean of reparation.

Climate justice also means that workers in high-carbon sectors, many of whom have sacrificed their health in coal mines and oil refineries, must be full and democratic participants in this justice-based transition. The guiding principle must be: No worker left behind.

Here are a couple of examples from my country. There is a group of oil workers in the Alberta tar sands who have started an organization called Iron and Earth. They are calling on our government to retrain laid-off oil workers and put them back to work installing solar panels, starting with public buildings like schools. It's an elegant idea, and almost everyone who hears about it supports it.

Our postal workers union, meanwhile, has been facing a push to shut down post offices, restrict mail delivery, and maybe even sell off the whole service to FedEx. Austerity as usual. But instead of fighting for the best deal they can get under this failed logic, they have put together a visionary plan for every post office in the country to become a hub for the green transition—a place where you can recharge electric vehicles and do an end-run around the big banks and get a loan to start an energy co-op; and where the entire delivery fleet is not only electric and made in Canada but also does more than deliver mail: It delivers locally grown produce and checks in on the elderly.

These are bottom-up, democratically conceived plans for a justice-based transition off fossil fuels. And we need them developed in every sector (from health care to education to media) and multiplied around the world.

Sounds pricey, you say? Good thing we live in a time of unprecedented private wealth. For starters, we can and must take the profits from the dying days of fossil fuels and spend them on climate justice. To subsidize free public transit and affordable renewable power. To help poor nations leapfrog over fossil fuels and go straight to renewables. To support migrants displaced from their lands by oil wars, bad trade deals, drought, and other worsening impacts of climate change, and by the poisoning of those lands by mining companies, many based in wealthy countries like yours and mine.

The bottom line is this: As we get clean, we have got to get fair. More than that, as we get clean, we can begin to redress the founding crimes of our nations: Land theft, genocide, slavery. Yes, the hardest stuff. Because we haven't just been procrastinating climate action all these years. We've been procrastinating and delaying the most basic demands of justice and reparation. And we are out of time on every front.

All this should be done because it's right and just, but also because it's smart. The hard truth is: Environmentalists can't win the emission-reduction fights on our own. It's not a slight against anyone; the lift is just too heavy. This transformation represents a revolution in how we live, work, and consume.

To win that kind of change, it will take powerful alliances with every arm of the progressive coalition: trade unions, migrant rights,

Indigenous rights, housing rights, transit, teachers, nurses, doctors, artists. To change everything, it takes everyone.

And to build that kind of coalition, it's got to be about justice: economic justice, racial justice, gender justice, migrant justice, historical justice. Not as afterthoughts but as animating principles. That will only happen when we take real leadership from those most impacted. Murrawah Johnson, an amazing young Indigenous leader who is at the heart of the struggle against the Carmichael mine, put it very well the other night here in Sydney: "People need to learn to be led."

Not because it's "politically correct," but because justice in the here and now is the only thing that has ever motivated popular movements to throw heart and soul into struggle. I'm not talking about going to a march or signing a petition, though there is a place for that. I'm talking about the sustained, daily, and long-haul work of social transformation. It's the thirst for justice—the desperate bodily *need* for justice—that builds movements like that.

We need warriors in this fight, and warriors don't step up *against* the accumulation of carbon in the atmosphere, not on its own, anyway. Warriors step up *for* the right to clean water, to good schools, to desperately needed decent-paying jobs, to universal health care. Warriors step up for the reunification of families separated by war and cruel immigration policies.

You already know that there will be no peace without justice—that's the core principle of the Sydney Peace Foundation. But here is what we need to understand just as well: There is no climate change breakthrough without justice, either.

Perhaps I should apologize for this kind of battle talk at an event celebrating peace. But we have to be clear that this is a fight, one in desperate need of a warrior spirit. Because as much as humanity has to win in this battle, the fossil fuel companies have a hell of a lot to lose. Trillions in income represented by all that unburnable carbon. Carbon in their current reserves and in the new reserves they are spending tens of billions to search out every year.

And the politicians who have thrown their lot in with these interests have a great deal to lose, too. Campaign donations, sure. The benefit of that revolving door between elected office and the extractive sector, too. But maybe most important, the money that comes when you don't have to think or plan—just dig. Right now, Australia is getting windfall profits from exporting coal to China. It's not the only way to fill government coffers, but it's most certainly the laziest: no pesky industrial planning, no tax or royalty increases on the corporations and billionaires with the resources to buy limitless attack ads.

All you have to do is hand out the permits, roll back some environmental laws, put new draconian restrictions on protest, call legitimate court challenges "green lawfare," trash the greenies nonstop in the Murdoch press, and you are good to go.

Which is why we shouldn't be surprised by the scathing assessment offered just last month by Michael Forst, the UN Special Rapporteur on the situation of human rights defenders. After a visit to Australia, he wrote:

> I was astonished to observe mounting evidence of a range of accumulative measures that have levied enormous pressure on Australian civil society. . . . I was astounded to observe what has become frequent public vilification of rights defenders by senior government officials, in a seeming attempt to discredit, intimidate and discourage them from their legitimate work.

It *is* striking that many of the people doing the most crucial work in this country—protecting the most vulnerable people and defending fragile ecologies from industrial onslaught—are facing a kind of dirty war. And we know all too well that it doesn't take much for this kind of political and media war to turn into a physical war, with very real casualties.

We see it around the world when land defenders try to stop mines, deforestation, and mega dams, from Honduras to Brazil. We see it when communities in India and the Philippines have tried to stop coal power stations because they are a threat to their water and wetlands. Not a metaphorical war, but real war, with lethal live ammunition fired into the bodies of the people getting in the way of the bulldozers.

According to Global Witness, this worldwide war is getting worse: They report that "More than three people were killed a week in 2015 defending their land, forests and rivers against destructive industries. . . . These numbers are shocking, and evidence that the environment is emerging as a new battleground for human rights. Across the world industry is pushing ever deeper into new territory. . . . Increasingly communities that take a stand are finding themselves in the firing line of companies' private security, state forces and a thriving market for contract killers." About 40 percent of the victims, they estimate, are Indigenous.*

And let us not tell ourselves that this happens only in so-called developing nations. We are seeing the war for the planet escalate right now in the United States, in North Dakota, where police and private security who look like they stepped off the battlefield

*This war has entered a new, more lethal phase with the election of Jair Bolsonaro as president of Brazil. Bolsonaro has made opening up the Amazon to unfettered development a top priority and has attacked Indigenous land rights, declaring ominously that "We're going to give a rifle and a carry permit to every farmer."

in Fallujah brutally repress a nonviolent Indigenous movement of water protectors.

The Standing Rock Sioux are trying to stop an oil pipeline that poses a very real threat to their water supply and, if built, would help hurtle us toward planet-destabilizing warming. For this, unarmed land defenders have been shot with rubber bullets, sprayed with pepper spray and other gases, blasted with sound cannons, attacked by dogs, put in what have been described as dog kennels, strip-searched, and arrested.

My fear is that the vilification of land defenders that we are seeing here in Australia—all the various and overlapping attempts at delegitimization, layered on top of openly racist portrayals of Indigenous people in the media, coupled with an increasingly draconian security state—prepares the ground for attacks like these.

So, though I continue to feel queasy about the carbon I burned on the flight, I am more than happy to be here, if only to play the role of the confused foreign meddler, the one saying, "Hold up a minute. We know where this leads, and this is a dangerous path you are going down." This beautiful and beautifully diverse country deserves better.

Oh, and this idea that your coal is somehow a humanitarian gift to India's poor? That has to stop. India is suffering more under coal pollution and the climate change it fuels than almost anywhere else on earth. A few months ago, it was so hot in Delhi that some of the roads melted. Since 2013, more than four thousand Indians have died in heat waves. This week, they closed all the schools in Delhi because pollution was so thick that they had to declare an emergency.

Meanwhile, the price of solar has plummeted by 90 percent and is now a more viable option for electrification than coal, especially because it requires less infrastructure and lends itself so well to community control. Many communities are demanding it, but in India, as elsewhere, the biggest barrier is the nexus of Big Government and Big Carbon: When people can generate their own electricity from panels on their rooftops, and even feed that power back into a micro grid, they are no longer customers of giant utilities; they are competitors. No wonder so many roadblocks are being put up: Corporations love nothing more than a captive market.

It is this cozy setup that the Indigenous rights and climate justice movement threatens to upend. And upend it we will—but let us be clear as we celebrate peace that it's going to be the fight of our lives.

SEASON OF SMOKE

It begins to strike me how precarious it all is, this business of not being on fire.

SEPTEMBER 2017

THE NEWS FROM THE NATURAL WORLD THESE DAYS IS MOSTLY ABOUT WATER, and understandably so.

We hear about the record-setting amounts of water that Hurricane Harvey dumped on Houston and other Gulf cities and towns mixing with petrochemicals to pollute and poison on an unfathomable scale. We hear, too, about the epic floods that have displaced hundreds of thousands of people from Bangladesh to Nigeria (though we don't hear enough). And we are witnessing, yet again, the fearsome force of water and wind as Hurricane Irma, one of the most powerful storms ever recorded, leaves devastation behind in the Caribbean, with Florida now in its sights.

Yet, for large parts of North America, Europe, and Africa, this summer has not been about water at all. In fact, it has been about its absence; it's been about land so dry and heat so oppressive that forested mountains exploded into smoke like volcanoes. It's been about fires fierce enough to jump the Columbia River; fast enough to light up the outskirts of Los Angeles like an invading army; and pervasive enough to threaten natural treasures like the tallest and most ancient sequoia trees and Glacier National Park.

For millions of people from California to Greenland, Oregon to Portugal, British Columbia to Montana, Siberia to South Africa, the summer of 2017 has been the summer of fire. And more than anything else, it's been the summer of ubiquitous, inescapable smoke.

For years, climate scientists have warned us that a warming world is an extreme world, one in which humanity is buffeted by both brutalizing excesses and stifling absences of the core elements that have kept fragile life in equilibrium for millennia. At the end of the summer of 2017, with major cities submerged in water and others licked by flames, we are currently living through Exhibit A of this extreme world, one in which natural extremes come head-to-head with social, racial, and economic ones.

#FAKEWEATHER

I checked the forecast before coming to British Columbia's Sunshine Coast, a ragged strip of coastline marked by dark evergreen forests that butt up against rocky cliffs and beaches strewn with driftwood, the charming flotsam from decades of sloppy logging operations. Reachable only by ferry or floatplane, this is the part of the world where my parents live, where my son was born, and where my grandparents are buried. Though it still feels like home, we now get here for only a few weeks a year.

The government of Canada weather site predicted that the next week would be glorious: an uninterrupted block of sun, clear skies, and higher-than-average temperatures. I pictured hot afternoons paddling in the Pacific and still, starry nights.

But when we arrive in early August, a murky blanket of white has engulfed the coast, and the temperature is cool enough for a sweater. Forecasts are often wrong, but this is more complicated. Somewhere up there, above the muck, the sky *is* clear of clouds. The sun *is* particularly hot. Yet intervening in those truths is a factor the forecasters did not account for: huge quantities of smoke, blown up to four hundred miles from the province's interior, where about 130 wildfires are burning out of control.

Enough smoke has descended to turn the sky from periwinkle blue to this low, unbroken white. Enough smoke to reflect a good portion of the sun's heat back into space, artificially pushing temperatures down. Enough smoke to transform the sun itself into an angry pinpoint of red fire surrounded by a strange halo, unable to burn through the relentless haze. Enough smoke to blot out the stars. Enough smoke to absorb any possible sunsets. At the end of the day, the red ball abruptly disappears, only to be replaced by a strange burnt-orange moon.

The smoke has created its own weather system, one powerful enough to transform the climate not just where we are, but in a stretch of territory that appears to cover roughly a hundred thousand square miles. And the smoke, a giant smudge on the satellite images, respects no borders: Not only is about a third of British Columbia choked, but so are large parts of the Pacific Northwest, including Seattle, Bellingham, and Portland, Oregon. In the age of #FakeNews, this is #FakeWeather, a mess in the sky created, in large part, by toxic ignorance and political malpractice.

Up and down the coast, the government has issued air quality warnings, urging people to avoid strenuous activity. Beyond a certain threshold, fine particulate matter in the air is officially unsafe, bad enough to cause health problems. The air in parts of Vancouver is three times above that safe threshold, with some smaller communities on the coast significantly worse off. Elderly people and other sensitive populations are being urged to stay inside—or, better yet, to go somewhere with a decent air-filtration system. One local official recommends a trip to the mall.

INLAND INFERNOS

At the epicenter of the disaster, where the flames are closing in, the air quality is far worse. Anything over 25 micrograms of fine particulates per cubic meter is considered unsafe. Kamloops, the city currently housing many of the evacuees, averages 684.5 micrograms per cubic meter. That rivals Beijing on some of its very worst days. Airlines cancel flights, and people suffering from breathing problems pack emergency rooms.

Since this disaster began, some 840 separate fires have ignited, forcing, at this point, some 50,000 people to evacuate their homes, according to the Red Cross. In early July, the government declared a rare state of emergency and by the time we arrive, it has already been extended twice. Hundreds of structures have been razed, and some whole communities, including Indigenous reserves, have been reduced mostly to ash.

So far, roughly 1,800 square miles of forest, farm, and grassland have burned. That makes this the second-largest fire disaster in British Columbia's history—and it's still going strong, putting the all-time record within grasp.

I call a friend in Kamloops. "Everyone who can is taking their kids far away, especially little ones."

Which puts things into perspective for us on the coast. It may be smoky, but we're damn lucky.

IT WILL BLOW OVER

Since the New Year, and the new US administration, I haven't taken a day off, let alone a weekend. Like so many others, I've attended way too many meetings and marched until my feet blistered. I wrote a book in a blur, then toured with it. And my husband, Avi, and I helped start The Leap, a new political organization. Throughout the winter and spring, "B.C. in August" was our family mantra. It was the finish line (albeit a temporary one), and we fully planned to collapse on it. It was also the way we kept our five-year-old son, Toma, in the game. On cold nights in the east, we mapped out the forested walks we would take, the canoe trips, the swims. We imagined the blackberries we would pick, the crumbles we would bake; we listed the grandparents, aunts, uncles, cousins, and old friends we would visit.

This break ("self-care" in the parlance of my younger coworkers) took on mythic qualities in our home. Which may be why I am a bit slow to clue in to the seriousness of the fires—and the smoke.

On the first day, I'm sure the sun will burn it away by noon. By evening, I announce that it will blow over by morning, revealing at least a glimpse of actual sky. For the first week, I greet each day hopefully, convinced that the drab light peeking through the curtains is just morning mist. Every day, I am wrong.

The placid weather forecast that seemed so promising before we traveled turns out to be a curse. Sunny, windless days mean that the smoke, once it is upon us, parks over our heads like an unmoveable outdoor ceiling. Day after day after day.

My allergies are going nuts. I bathe my eyes in drops and pop antihistamines well beyond the recommended dosage. Toma breaks out in hives so severe he needs steroids.

I keep taking my glasses off and cleaning them, rubbing them first with my shirt, then a microfiber cloth, then proper glass cleaner. Nothing helps. Nothing makes the smudge disappear.

MISSING BLUE

A week into the whiteout, the world begins to feel small. Life beyond the smoke starts to seem like a rumor. At the ocean's edge, we can usually look across the Salish Sea to Vancouver Island; now we strain to see an outcropping of rock a few hundred feet from shore.

I've been on this coast for whole winters when we barely saw the sun. I learned to love the steely beauty, the infinite shades of gray chiseled in the mountains. The low sky and the movement of the mist. But this is different. There's a lifeless quality to the smoke; it just sits there, motionless and monotone.

Blankets of smog are something a lot of people on this planet have learned to live with in big polluted cities such as Beijing, New Delhi, São Paulo, and Los Angeles. Smoke from wildfires is a little different. In part because you know that you are not breathing pollution from power plants or exhaust from cars but, rather, smoke from trees that were very recently alive. You are breathing in forest.

I decide that the animals are depressed. The seals seem to pop their heads up in a purely utilitarian fashion, just to take a breath and then disappear again beneath the gray surface. They do not play. The eagles, I am convinced, are flying for function, not fun—no soaring or wind surfing. There's little doubt I'm imagining all this, projecting, anthropomorphizing—it's a bad habit.

I email a friend in Seattle, a prominent environmentalist, to ask him how he is faring in the smoke. He reports that the birds have stopped singing, and he is mad all the time. At least I'm not the only one.

WHAT IF WE'RE NEXT?

It begins to strike me how precarious it all is, this business of not being on fire.

This part of British Columbia, technically a temperate rain forest, is a tinderbox. So far this summer, less than half an inch of rain has fallen. The forest ground cover, usually moist and squishy, is yellow and desiccated and crunches underfoot. You can smell the flammability.

The roads are lined with yellow signs announcing a total ban on open fires. Every time we turn on the radio, we hear warnings, increasingly frantic, about open fires, cigarettes tossed out of cars, and fireworks. One guy earned himself a night in jail and over a thousand dollars in fines after he drunkenly celebrated the fact that his home had survived a brush with fire by setting off fireworks—which could well have set yet another blaze.

It's clear that one lightning storm, or a couple of careless campers, would be enough to send this place up. We've come close before. Two years ago, a serious wildfire threatened part of the coast about twenty minutes from here, taking the life of a local man who was helping to fight the flames. Yet, despite the years I have spent living here, until this week, I've never really thought about what it would mean if a fire like that ever got out of control. Now I do, and it's unsettling. The Sunshine Coast has a year-round population of thirty thousand people served by a single highway that ends in a ferry dock. What the hell does an emergency evacuation look like in a place with no roads out?

I ask local friends. They look worried and talk about who has which kind of fishing boat.

A DEATH IN A BLUEBERRY FIELD

Nine days into the whiteout, some terrible news arrives. A farmworker in smoke-choked Sumas, Washington (less than a mile from the Canadian border) has died in a Seattle hospital. Honesto Silva Ibarra came to the United States from Mexico on a temporary H-2A visa to work through the harvest season. He was twenty-eight years old and had been picking blueberries at Sarbanand Farms, owned by California-based Munger Farms, when he started feeling sick.

Silva's coworkers blame his death on unsafe working conditions: long hours, few breaks, insufficient food and drinking water—all compounded by the heavy smoke drifting in from British Columbia. "The workers have been overworked, underfed, have not been hydrated enough, and this has been going on for weeks," said Rosalinda Guillen, director of the advocacy group Community to Community Development. Some workers had fainted on the job, they told reporters.

A representative for Munger Farms says that Silva died after running out of his diabetes medication and that heat and wildfire smoke had "nothing to do" with it. The company also claims it did all it could to save him.

Whatever the cause (or causes) of death, the way the company treated Silva's coworkers when they raised their complaints is a chilling window into just how precarious life can be for America's thousands of guest workers. After Silva was hospitalized, workers staged a one-day strike to demand answers and better conditions. Sixty-six of them were immediately fired for insubordination. They found themselves without means to get home to Mexico and without payment for their final days of work. After setting up a protest camp, marching to the company's offices, and attracting local media, the workers won their back pay, and Munger has "voluntarily offered to provide safe transportation home for all of the terminated workers," according to the company spokesperson.

But they did not get their much-needed jobs back. Munger supplies Walmart, Whole Foods, Safeway, and Costco.

North of the border, there are similar reports of temporary farmworkers fainting and becoming sick on the job, with smoke apparently playing a role. And advocates point out that rather than being looked after, their sponsoring employers often send sick workers home like defective goods. According to the Canadian Broadcasting Corporation, at least ten workers in hot and smoky British Columbia were sent back to Mexico and Guatemala, "deemed too ill to work."

A HIGHLY DIVIDED DISASTER, AS USUAL

We learn the same lesson over and over again: In highly unequal societies, with deep injustices reliably tracing racial fault lines, disasters don't bring us all together in one fuzzy human family. They take preexisting divides and deepen them further, so the people who were already getting most screwed over before the disaster get extra doses of pain during and after.

We know a fair bit about how that looks during storms like Katrina, Sandy, Harvey, and Irma. We understand less about fire. But that's changing. We know, for instance, that as California struggles with a now-endless fire season, the state has become intensely reliant on prison labor, with inmates paid a staggeringly low hourly rate of one dollar to do some of the most dangerous work fighting flames. We know that hundreds of South African workers were contracted to help battle Alberta's 2016 Fort McMurray fire—only for them to stop working en masse after discovering that they were being paid significantly less than their Canadian counterparts, and less than press reports claimed they were being paid. They were promptly sent home.

We also know, as with floods, that our media give far more coverage to house pets rescued from wildfires in the United States and Canada than to the human lives lost because of infernos in, say, Indonesia and Chile. A global 2012 study estimated that more than three hundred thousand people die annually as a result of the smoke and air pollution from wildfires, mostly in sub-Saharan Africa and Southeast Asia.

And in British Columbia this summer, we learned still more about the way inequalities play out against a burning backdrop. Several Indigenous leaders raised concerns that during fire emergencies, their communities do not receive the same level of urgent response as nonindigenous ones, whether for fighting the flames or for rebuilding afterward. With this in mind, several Indigenous reserves directly threatened by fire refused to evacuate, and a portion stayed behind to fight the flames—some with their own teams of trained firefighters and equipment, others with little more than garden hoses and sprinklers. In at least one case, the police responded by threatening to come in and remove children from their families, words with traumatic reverberations in a country that not so long ago systematically took Indigenous children from their homes as a matter of policy.

In the end, no First Nations homes were raided, and many were saved because of self-organized fire brigades. Ryan Day, chief of the fire-threatened Bonaparte Indian Band, said, "If we all evacuated, we would have no houses on this reserve."

A WORLD WITH TWO SUNS

It's almost a week into the smoke-out, and the moon is nearing full. Around here, people take full moons seriously; there are drug-fueled dance parties in the forest and late-night kayak excursions taking advantage of the extra light.

But when the almost-full moon rises in early August, I mistake it, at first, for the sun: It's the same shape and almost the same fiery color.

For about four days, it's as if we are on a different planet, one with two red suns and no moon at all.

SOUR FRUIT

It's week two of the smoke-out, and the blackberries are finally ripe. We set out to collect them. It feels strange to be going through with this carefree summer ritual with the air so thick and the news so grim, but we do it anyway. Combining hiking with nonstop eating is one of Toma's all-time favorite activities.

It's pretty much a bust. With so little rain and such a weak sun to warm them, even the ripest berries are sour. Toma quickly loses interest and refuses to try any more. We come home with shin scratches and an empty bucket.

We don't stop hiking, though. We spend hours every day walking through the stands of moss-covered cedars and Douglas firs, breathing in the super-oxygenated air. I love these forests and have never taken their primordial beauty for granted. Now I find myself in near worship—thanking them not just for scrubbing the air, or for the shade and the carbon sequestration they provide ("ecosystem services" in the lingo of business environmentalism), but for their sheer stamina. For not joining their burning brethren. For sticking with us, despite our failings. At least so far.

HELLO AGAIN

I have breathed this smoke before. Not these precise airborne particles, of course, but smoke from many of the same wildfires. And the odd thing is, I breathed it in some 570 miles east of here, in another province entirely.

I spent mid-July in Alberta, helping teach a course on environmental reporting at the Banff Centre for Arts and Creativity.

This time, too, the forecast had looked perfect: sunny, clear, warm. This time, too, the forecast was usurped from the first day by a smoke cast, a haze that obscured the spectacular mountains in Banff National Park and provoked air quality warnings, headaches, and a catch in the throat. More #FakeWeather.

Back in July, the winds were blowing east, which is why the Rockies were getting a face full of smoke. In Calgary, Canada's oil capital, the smoke was so thick that it obscured the city's skyline of gleaming glass towers bearing the logos of Shell, BP, Suncor, and TransCanada. And the smoke didn't stop there. It kept traveling eastward, reaching well into the center of the continent, to Saskatchewan and Manitoba and down to North Dakota and Montana. (NASA released a striking picture of the five-hundred-mile-long plume.)

Then, just as my family was heading to coastal British Columbia, the winds abruptly shifted and started blowing the plume westward, with the Rocky Mountains now acting like a giant tennis racket, lobbing the smoke to the Pacific.

Inhaling smoke originating from the same incinerated forests for the second time in one summer—never mind that I had traveled six hundred miles and crossed a provincial boundary—was an eerie experience. I felt like the smudge was somehow stalking me like the smoke monster on *Lost*.

WORLD ON FIRE

Part of the head trip in all this is the sheer scale of the disaster, both temporally and spatially. Even devastating hurricanes like Harvey tend to concentrate their impacts in a contained geography. And the event (though not the aftermath) is relatively brief.

These fires, which rage for months, are of a different order entirely. There are the direct impacts of the fires. The huge swath of land charred. The tens of thousands of lives overturned by evacuation orders. The lost homes and farms and cattle. The industries (from tourism operators to sawmills) forced to close down.

And then there are the less direct impacts of all that wandering smoke. Over July and August, the smoke from this conflagration covered an area spanning roughly seven hundred thousand square miles. That's bigger than all of France, Germany, Italy, Spain, and Portugal combined. All touched by this one fast-moving disaster.

And this is just one snapshot of a much larger season of fire. At summer's end, large parts of the American West were on fire. A fire in Los Angeles was the largest ever recorded within city limits; a fire-related state of emergency was declared for every single county in Washington State. In Montana, a wildfire named the Lodgepole Complex burned some 425 square miles of territory, making it the third-largest blaze in the recorded history of the region. This is part of a broader increase in both the numbers of fires and the months during which they burn: Since the 1970s, fire season in the United States has lengthened by 105 days, according to an analysis by Climate Central.

The area of Europe that has burned this fire season has been triple the average, and it's not over yet. Central Portugal experienced the deadliest impacts: In June, more than sixty people died in a blaze near Pedrógão Grande.* Hundreds of homes have burned in Siberia. During Chile's summer months, the country battled the largest wildfire in its recorded history, and thousands of people were displaced. In June in South Africa, the same storm caused flooding in Cape Town and fanned the flames of unprecedented deadly wildfires in nearby towns. Even Greenland, that icy place, saw large and unusual wildfires this summer. Jason Box, a world-renowned climate scientist specializing in the Greenland ice sheet, pointed out that "temperatures in Greenland are probably higher [than they have been over] the last 800 years."

YES, IT'S CLIMATE CHANGE

Warmer and drier weather is not the only factor at play. Another is the perennially hubristic attempt to reengineer natural forces far more powerful than we are. Fire is a crucial part of the forest cycle: Left to their own devices, forests burn periodically, clearing the way for new growth and reducing the amount of highly flammable underbrush and old wood ("fuel" in firefighter parlance). Many Indigenous cultures have long used fire as a key part of land care. But

*That death toll was surpassed just one year later in neighboring Greece. Some one hundred people died in a series of fires that ripped through coastal lands starting in Attica, the deadliest fire in modern European history. The remains of a large family were found by a cliff's edge with their limbs entwined: they had clung to each other as the flames approached. "Instinctively, seeing the end nearing, they embraced," the head of Greece's Red Cross, Nikos Economopoulos, told a television crew.

in North America, modern forest management has systematically supressed cyclical fires in order to protect profitable trees that were headed for sawmills, and out of fear that small fires could spread to inhabited areas (and there are more and more inhabited areas).

Without regular natural burns, forests are chock-full of fuel, provoking fires to burn out of control. And there's a hell of a lot more fuel as a result of bark beetle infestations, which have left behind huge stands of dry and brittle dead trees. There is compelling evidence that the bark beetle epidemic has been exacerbated by climate change–related heat and drought.

Overlying it all is the uncomplicated fact that hotter, drier weather (which is directly linked to climate change) creates the optimal conditions for wildfires. Indeed, these forces have conspired to turn forests into perfectly laid campfires, with the dry earth acting like balled-up newspaper, the dead trees serving as kindling, and the added heat providing the match. Mike Flannigan, a University of Alberta wildfire expert, is blunt. "The increase in area burned in Canada is a direct result of human-caused climate change. Individual events get a little more tricky to connect, but the area burned has doubled in Canada since the 1970s as a result of warming temperatures." And according to a 2010 study, fire occurrence in Canada is projected to increase by 75 percent by the end of the century.

Here's the really alarming thing: 2017 was not even an El Niño year, the cyclical natural warming phenomenon that was commonly cited as a key factor in the huge fires that raged in Southern California and Northern Alberta last year.

With no El Niño to blame, some media outlets have been willing to drop the hedging. To quote Germany's Deutsche Welle: "Climate change sets the world on fire."

FAIRY TALES AND FEEDBACK LOOPS

"Looks like snow is coming," Toma declares solemnly, his face pressed up to the window and the white, thick air on the other side.

Ever since we left Alberta, his five-year-old mind has been struggling to understand the smoke that has defined his summer. Trying to make sense of my chronic cough and his raging skin rash. Struggling, most of all, with the soundtrack of worried chatter among the grown-ups in his life.

His response goes through phases: Nightmares wake him up at night. He writes songs with lyrics like "Why is everything going wrong?" There's a lot of inappropriate laughter.

At first, he was excited by the idea of wildfires, confusing them with campfires and angling for s'mores. Then his grandfather explained that the sun had turned into that weird, glowing dot because the forest itself was on fire. He was stricken. "What about the animals?"

We have developed techniques for controlling worry. They begin with taking deep breaths, and we do it several times a day. But it occurs to me that breathing extra amounts of this particular air is probably not great, especially for small lungs already prone to infection.

Avi and I don't talk to Toma about climate change, which may seem strange given that I write books about it and Avi directs films about it, and we both spend most of our waking hours focused on the need for a transformative response to the crisis. What we do talk about is pollution, though on a scale Toma can understand. Like plastic, and why we have to pick it up and use less of it because it makes the animals sick. Or we look at the exhaust coming out of cars and trucks and talk about how you can get power from the sun and the wind and store it in batteries. A little kid can grasp concepts like these and know exactly what should happen (better than plenty of adults). But the idea that the entire planet has a fever that could get so high that much of life on Earth could be lost in the convulsions—that seems to me too great a burden to ask small children to carry.

This summer marks the end of his protection. It isn't a decision I'm proud of, or one I even remember making. He just heard too many adults obsessing over the strange sky and the real reasons behind the fires, and he finally put it all together.

At a playground in the haze, I meet a young mother who offers advice on how to reassure worried kids. She tells hers that forest fires are a positive part of the cycle of ecosystem renewal—the burning makes way for new growth, which feeds the bears and deer.

I nod, feeling like a failed mom. But I also know that she's lying. It's true that fire is a natural part of the life cycle, but the fires currently blotting out the sun in the Pacific Northwest are the opposite—they're part of a planetary death spiral. Many are so hot and intransigent that they are leaving scorched earth behind.* The rivers of bright red fire retardant being sprayed from planes are seeping into waterways, posing a threat to fish. And just as my son fears, animals are losing their forested homes.

The biggest danger, however, is the carbon being released as the forests burn. Three weeks after the smoke descended on the coast, we learn that the total annual greenhouse gas emissions for the province of British Columbia had tripled as a result of the fires, and it's still going up.

This dramatic increase in emissions is part of what climate scientists mean when they warn about feedback loops: burning

*As was the case in November 2018 in Paradise, California, a town of 27,000 people that was razed to the ground in the deadliest fire in the state's history.

carbon leads to warmer temperatures and long periods without rain, which leads to more fires, which release more carbon into the atmosphere, which leads to even warmer and drier conditions, and even more fires.

Another such lethal feedback loop is playing out with Greenland's wildfires. Fires produce black soot (also known as "black carbon"), which settles on ice sheets, turning the ice gray or black. Darkened ice absorbs more heat than reflective white ice, which makes the ice melt faster, which leads to sea level rise and the release of huge amounts of methane, which causes more warming and more fires, which in turn create more blackened ice and more melting.

So, no, I'm not going to tell Toma that the fires are a happy part of the cycle of life. We settle for half truths and fudging to make the nightmare subside. "The animals know how to escape from the fires. They run to rivers and streams and other forests."

We talk about how we need to plant more trees for the animals to come home to. It helps, a little.

A WAKE-UP CALL—FOR SOME

One of the regions hit hardest by the fires is a place I have visited often, the territory of the Secwepemc people, which encompasses a huge swath of land in British Columbia's interior—much of it now on fire. The late Arthur Manuel, a former Secwepemc chief, was a dear friend and hosted me several times. So far in 2017, I have visited his territory twice: once to attend Manuel's funeral and once for a meeting he had been organizing when heart failure took his life.

The gathering was in response to Canadian prime minister Justin Trudeau's decision to approve a $7.4 billion project that would nearly triple the capacity of the Kinder Morgan Trans Mountain Pipeline, which carries high-carbon tar sands oil from Alberta through British Columbia. The expanded network of pipes would pass through dozens of waterways on Secwepemc land, and is forcefully opposed by many traditional landholders. Arthur believed the struggle has the potential to turn into "Standing Rock North."

When the fires began this summer, Manuel's friends and family wasted no time making the argument that building more fossil fuel infrastructure as the world burns is both absurd and reckless. A statement was issued by the Secwepemc Working Group on Indigenous Food Sovereignty opposing the pipeline expansion project and demanding that the existing, smaller pipeline be shut down immediately to reduce the risk of a catastrophic accident should fire and oil meet.

"We are in a critical state of emergency dealing with the impacts of climate change," said Secwepemc teacher Dawn Morrison. "The health of our families and communities relies heavily on our ability to harvest wild salmon and access clean drinking water, both of which are at risk if the Kinder Morgan pipeline was ruptured or impacted by the fires."

This is common sense: When oil and gas infrastructure finds itself in the bull's-eye of the cumulative effects of burning so much fossil fuel—think of oil rigs battered by superstorms, or Houston underwater—we should all do what the Secwepemc did: treat the disaster as a wake-up call about the need to build a safer society, fast.

WHATEVER YOU DO, DON'T TALK ABOUT OIL

Our political and economic systems, however, are not built that way; indeed, they are built to actively override that kind of survival response. So Kinder Morgan doesn't even bother answering the community's concerns. What's more, the company is gearing up to begin construction on the expansion this month, with the fires still raging.

Worse, some extractive industries are actively using the fiery state of emergency to get stuff done that was impossible during normal times. For instance, Taseko Mines has been fighting for years to build a highly contentious, open-pit gold and copper mine in one of the parts of British Columbia hit hardest by the fires. Fierce opposition among the Tsilhqot'in First Nation has so far successfully fended off the toxic project, resulting in several key regulatory victories.

But this July, with several of the impacted Tsilhqot'in communities under evacuation order or holding their ground to fight the fires themselves, the outgoing British Columbia government, notorious as a "Wild West" of political payola, did something extraordinary. In its last week in office after suffering a humiliating election defeat, the government handed Taseko a raft of permits to move ahead with exploration. "It defies compassion that while our people are fighting for our homes and lives, B.C. issues permits that will destroy more of our land beyond repair," said Russell Myers Ross, a Tsilhqot'in chief. A representative of the outgoing government responded: "I appreciate this may come at a difficult time for you given the wildfire situation affecting some of your communities." Despite the stresses the fires have placed on their people, the Tsilhqot'in are fighting the move in court, and the company has already been forced to suspend its drilling plans in the face of legal troubles.

Yet anyone holding out hope that the fires might jolt Prime Minister Justin Trudeau into serious climate action has been gravely disappointed. Canada's prime minister loves being photographed frolicking in British Columbia's spectacular wilderness (preferably shirtless), and his wife, Sophie Grégoire, recently unleashed a hurricane of emojis by posting a picture of herself surfing off Vancouver Island. (It was during the fires and the sky looked hazy.)

But for all his gushing about British Columbia's forests and coastal waters, Trudeau is slamming his foot on the accelerator when it comes to pipelines and tar sands expansion. "No country would find 173 billion barrels of oil in the ground and just leave them there," he told a cheering crowd of oil and gas executives in Houston in March 2017. He hasn't budged since. Never mind that Houston has since been flooded by an unprecedented storm, or that a third of Trudeau's own country is on fire. This month, one of his top ministers said of the Kinder Morgan pipeline approval, "Nothing that's happened since then has changed our mind that this is a good decision." Trudeau is on fossil fuel autopilot, and nothing, it seems, will make him swerve.

Then there is President Donald Trump, whose climate crimes are too comprehensive and too layered to delineate here. It does seem worth mentioning, however, that he chose this summer of floods and fires to disband the federal advisory panel assessing the impacts of climate change on the United States and to greenlight Arctic drilling in the Beaufort Sea.*

*One year later, in 2018, Trump's (now former) interior secretary used California's record-breaking wildfires to quietly open up large new tracts of forest for logging. This was nothing new for Ryan Zinke, who in 2015, while serving as a member of Congress, cosponsored legislation that threatened environmental protection of public forests. Three years later, he was still repeating the mantra that deforestation was the best option to curb forest fires. "Every year we watch our forests burn, and every year there is a call for action. Yet, when action comes, and we try to thin forests of dead and dying timber, or we try to sustainably harvest timber from dense and fire-prone areas, we are attacked

THE GUY WHO LOST TWO HOUSES

It's not just politicians who are bound and determined not to learn from nature's blaring messages.

In the midst of British Columbia's fire emergency, the Canadian Broadcasting Corporation struck human interest gold: They found a man, Jason Schurman, whose log home burned down in British Columbia—and who also lost a home in the Fort McMurray fires one year earlier. Two homes, two fires, same guy. The CBC ran photos of his charred properties (separated by eight hundred miles) side by side. In both cases, only the fireplace and chimney remained.

There's a lot of moving detail in the story about the human wreckage these disasters leave behind: the endless paperwork, traumatic memories, and family stress. But there is no mention of climate change. This is notable because Schurman works as a site supervisor in the Alberta tar sands. Still, the reporter did not ask Schurman if losing two homes and nearly losing his son raised any questions for him about the industry in which he works (one of the only sectors in Canada or the United States still paying blue-collar salaries that afford middle-class lives). Instead, the tale of a man "twice burned" played as a quirky human interest story, alongside one about a firefighter who got married amid the flames.

When *Vice* picked up the irresistible story, the reporter did raise climate change with Schurman, which he acknowledged

---

with frivolous litigation from radical environmentalists who would rather see forests and communities burn than see a logger in the woods." There is no doubt that California needs both better forest management and wiser land use policy. But given the indispensable role that trees play in keeping carbon out of the atmosphere, the last thing we should be doing is expanding deforestation in the name of fire prevention.

could be one of the factors contributing to these infernos. But, in very *Vice*-like fashion, the bulk of the article focused on how the oil worker is coping with his losses through baroque body art: "The constant pain of a tattoo also takes your mind completely off . . . losing everything I own."

YOU GET USED TO IT

Aren't we all guilty, in one way or another, of sleepwalking toward apocalypse? The soft-focus quality the smoke casts over life here seems to make this collective denial more acute. Here on the coast in August, we all look like sleepwalkers, stumbling around doing our work and errands, having vacations in a thick cloud of smoke, pretending we don't hear the alarm clanging in the background.

Smoke, after all, is not fire. It's not a flood. It doesn't command your immediate attention or force you to flee. You can live with it, if less well. You get used to it.

And that's what we do.

We paddleboard in the smoke and act like it's mist. We bring beers and ciders to the beach and remark that, on the upside, you barely need sunscreen at all.

Sitting on the beach under that fake, milky sky, I suddenly flash to those images of families sunning themselves on oil-soaked beaches in the midst of the BP Deepwater Horizon disaster. And it hits me: They are us, refusing to let a raging, record-breaking wildfire interfere with our family vacation.

During disasters, you hear a lot of praise for human resilience. And we are a remarkably resilient species. But that's not always good. It seems a great many of us can get used to almost anything, even the steady annihilation of our own habitat.

A WINDOW TO A HACKED PLANET

A week into what a local Sunshine Coast paper has dubbed our "Days of Haze," The Atlantic runs a cheery story headlined, TO STOP GLOBAL WARMING, SHOULD HUMANITY DIM THE SKY?

The piece focuses on a method often referred to as solar radiation management, which would see sulfur dioxide sprayed into the stratosphere to create a barrier between Earth and the sun, forcibly lowering temperatures. Trump's withdrawal from the Paris Agreement, the piece notes, means more governments, including China, are taking sun-dimming seriously.

The first mention of possible risks comes in paragraph 20, where the journalist quotes a climate scientist saying that hacking the planet "could induce droughts or floods or things like that." Yeah, that would be a bummer. In fact, there is a large cache of peer-reviewed research showing that this form of geoengineering could interfere with the monsoons in Asia and Africa, thus threatening the food and water supply for billions of people.

Now imagine a scenario in which men like Trump, Indian prime minister Narendra Modi, and North Korea's "Supreme Leader" Kim Jong-un were empowered to deploy these climate-altering technologies like unconventional weapons, hurtling us into an era of undeclared weather wars in which one country sacrifices the precipitation of another in order to save its crops, and the other retaliates by unleashing mega-floods.

Some would-be planet hackers insist that these worst-case risks can be managed (though they never explain how). All concede to lesser downsides, however. Spraying sulfur dioxide into the stratosphere would almost certainly create a permanent milky-white haze, making clear blue skies a thing of the past for the entire planet. The haze might well prevent astronomers from seeing the stars and planets clearly, and weaker sunlight could reduce the capacity of solar power generators to produce energy.

When you think about this in the abstract, it can seem like a small price to pay to buy "some more time to get our act together" on pollution control, as the *Atlantic* article puts it. But it is something else entirely to read about the prospect of deliberately dimming the sky under a sky that is already artificially dimmed with omnipresent smoke casting a literal pall over daily life.

Losing the sky is no small thing. We take it for granted that even in the most crowded of cities, we can look up and see the world beyond our reach—yes, there are planes and satellites, but beyond that is the heavens, the unknown, the ultimate "out there." In almost every corner of the Pacific Northwest this August, when we gazed up at the sky, we didn't see any of that expanse. We just saw ourselves, more detritus of our own broken system. In the blanket of smoke, we had a ceiling, not a sky—and it felt like a suffocating lid placed over possibility itself.

I hear myself manically suggesting to Avi that we should just drive north until we hit clean air. And then I remember that if we did, we'd be face-to-face with rapidly melting permafrost. We stay put.

THE WIND CHANGES

Almost two full weeks into the smoke-out, something shifts. I hear it first, and then I see branches moving: wind. A sudden temperature drop. And by noon, actual patches of blue, separated by clouds. I had forgotten how distinct they are from haze—higher, for starters, and with all kinds of delicate shapes and movement.

The smoke hasn't cleared entirely, but enough of it has blown away to suddenly make the world look sharp. Crisp. You know that elation when a long fever finally breaks? I feel like that.

The next day brings rain; not a lot, but enough to hope for some relief for the 2,400 exhausted and overworked firefighters. My allergies clear up, and Toma starts sleeping through the night again.

But the news from the interior is disastrous. The same wind that finally pried loose the blanket of smoke on the coast has been fanning the flames at the epicenter of the fires. The stillness that trapped the smoke here for so long turns out to have been the only bright spot for the fire brigades. Now that's over, and there's not nearly enough rain.

Over the next week, British Columbia blazes through the record books. By mid-August, the fires break the provincial record for the most land burned in one year: 3,453 square miles.* Within days, several different fires combine forces to create the largest single fire in British Columbia's history.

TOO SOON

When the solar eclipse arrives, I feel nothing but dread. We have clear skies, a near-perfect viewing location, and I know, in theory, that what is about to happen is a natural wonder. But I'm just not ready to say good-bye to the sun again, even for a few minutes. We only just got it back.

I spend the eclipse sitting outside alone, staring at the horizon, clinging to the dying light. A week after neo-Nazis marched with torches in Charlottesville, Virginia, and with so much of the world engulfed in actual infernos, this sudden dimming of our world just feels too damn literal.

*That grim record was broken just one year later, during the historic fire season of 2018.

THE GLOBAL FIRE ALARM SYSTEM IS BROKEN

Over Labor Day weekend, more than 160 fires are still burning in British Columbia. Extremely hot, dry, and windy weather has conspired to create the conditions for a slew of large, new wildfires to ignite and for old ones to expand exponentially. Authorities announce new evacuations daily; at last count, some 60,000 displaced people had registered as evacuees with the Red Cross over the course of the summer. The state of emergency has been extended for the fourth time.

But even in Canada, it's impossible for this news to compete with the devastating fallout from Hurricane Harvey; the scores dead and millions impacted by record flooding in South Asia and Nigeria; and now the fury of Hurricane Irma. Then there are the headline-grabbing blazes in Los Angeles, the state of emergency in Washington State, and new evacuations ordered, from Glacier National Park to Northern Manitoba. A satellite image from early September shows the entire length of the continent blanketed in smoke, #FakeWeather from the Pacific to the storm-churning Atlantic.

I can barely keep track of the nonstop convulsions, and it's my job to do so. I do know this: Our collective house is on fire, with every alarm going off simultaneously, clanging desperately for our attention. Will we keep stumbling and wheezing through the low light, acting as if the emergency were not already upon us? Or will the warnings be enough to force many more of us to listen? To respond like the Secwepemc, who, in a cloud of smoke, are nonetheless putting their bodies on the line to stop an oil pipeline from being built on their fire-scarred land?

Those are the questions still hanging in the air at the end of this summer of smoke.

THE STAKES OF OUR HISTORICAL MOMENT

You went and showed us all that you can win. Now you have to win.

SEPTEMBER 2017
LABOUR PARTY CONFERENCE, BRIGHTON

IT'S BEEN SUCH A PRIVILEGE TO BE PART OF THIS HISTORIC CONVENTION. TO FEEL its energy and optimism.

Because, friends, it's bleak out there. How do I begin to describe a world upside down? From heads of state tweeting threats of nuclear annihilation, to whole regions rocked by climate chaos, to thousands of migrants drowning off the coasts of Europe, to openly racist parties gaining ground, most recently and alarmingly in Germany—most days there is simply too much to take in. So, I want to start with an example that might seem small against such a vast backdrop.

The Caribbean and southern United States are in the midst of an unprecedented hurricane season: pounded by storm after record-breaking storm. As we meet, Puerto Rico—hit by Irma, then Maria—is without power and could be for months. Its water and communication systems are also severely compromised. Three and a half million US citizens on that island are in desperate need of their government's help.

But just like during Hurricane Katrina, the cavalry is missing in action. Donald Trump is too busy trying to get black athletes fired, smearing them for daring to shine a spotlight on racist violence.

As if all this weren't enough, the vultures are now buzzing. The business press is filled with articles about how the only way for Puerto Rico to get the lights back on is to sell off its electricity utility. Maybe its roads and bridges, too.

This is a phenomenon I have called the shock doctrine, the exploitation of wrenching crises to smuggle through policies that devour the public sphere and further enrich a small elite. We see this dismal cycle repeat again and again. We saw it after the 2008 financial crash. We are already seeing it in how the Tories are planning to exploit Brexit to push through disastrous pro-corporate trade deals without debate.

The reason I am highlighting Puerto Rico is because the situation is so urgent. But also because it's a microcosm of a much larger global crisis, one that contains many of the same overlapping elements: accelerating climate chaos; militarism; histories of colonialism; a weak and neglected public sphere; a totally dysfunctional democracy. And overlaying it all: the seemingly bottomless capacity to discount the lives of huge numbers of black and brown people. Ours is an age when it is impossible to pry one crisis apart from all the others. They have all merged, reinforcing and deepening one another like one shambling, multiheaded beast. I think it's helpful to think of the current US president in much the same way.

It's tough to know how to adequately sum him up. So, let me try a local example. You know that horrible thing currently clogging up the London sewers, I believe you call it the "fatberg?" Well, Trump—he's the political equivalent of that: a merger of all that is noxious in the culture, economy, and body politic, all kind of glommed together in a self-adhesive mass. And we're finding it very, very hard to dislodge. It gets so grim that we have to laugh. But make no mistake: whether it's climate change or the nuclear threat, Trump represents a crisis that could echo through geologic time.

But here is my message to you today: Moments of crisis do not have to go the shock doctrine route—they do not need to become opportunities for the already obscenely wealthy to grab still more.

They can also go the opposite way.

They can be moments when we find our best selves, when we locate reserves of strength and focus that we never knew we had. We see it at the grassroots level every time disaster strikes. We all witnessed it in the aftermath of the Grenfell Tower catastrophe.* When the people responsible were MIA, the community came together, held one another in their care, organized the donations, and advocated for the living (as well as the dead). And they are doing it still, more than one hundred days after the fire, when there is still no justice and, scandalously, only a handful of survivors have been rehoused.

And it's not only at the grassroots level that we see disaster awaken something remarkable in us. There is also a long and proud history of crises sparking progressive transformation on a

*In June 2017, a fire broke out in a twenty-four-story public housing building in North Kensington, London, killing more than seventy people. Subsequent investigations found that various forms of neglect contributed to the building's vulnerability to the flames, including plastic-filled siding, which had been installed to improve the look of the tower's exterior but that proved to be hyperflammable; poorly maintained fire equipment; a broken ventilation system; and few escape routes.

society-wide scale. Think of the New Deal victories in the United States won by working people for social housing and old age pensions during the Great Depression. Or for the National Health Service in this country after the horrors of the Second World War.

This should remind us that moments of great crisis and peril do not necessarily need to knock us backward. They can also catapult us forward.

Our progressive ancestors achieved that at key moments in history. And we can do it again, in this moment when everything is on the line. But what we know from the Great Depression and the postwar period is that we never win these transformative victories by simply resisting, by simply saying no to the latest outrage.

To win in a moment of true crisis, we also need a bold and forward-looking "yes," a plan for how to rebuild and respond to the underlying causes of crisis. And that plan needs to be convincing, credible and, most of all, captivating. We have to help a weary and wary public to imagine itself into that better world.

And that is why I am so honored to be standing with you today. Because in the last election, that's exactly what the Labour Party did. Theresa May ran a cynical campaign based on exploiting fear and shock to grab more power for herself—first the fear of a bad Brexit deal, then the fear following the horrific terror attacks in Manchester and London. Your party and your leader, on the other hand, responded by focusing on root causes: a failed "war on terror," economic inequality, and weakened democracy.

But you did more than that.

You presented voters with a bold and detailed manifesto, one that laid out a plan for millions of people to have tangibly better lives: free tuition, fully funded health care, aggressive climate action. After decades of lowered expectations and asphyxiated political imagination, finally voters had something hopeful and exciting to say yes to. And so many of them did just that, upending the projections of the entire expert class.* You proved that the era of triangulation and tinkering is over. The public is hungry for deep change—they are crying out for it. The trouble is, in far too many countries, it's only the far right that is offering it, or seeming to, with that toxic combination of fake economic populism and very real racism.

You showed us another way, one that speaks the language of decency and fairness; that names the true forces most responsible for this mess, no matter how powerful; and that is unafraid of some of the ideas we were told were gone for good. Like wealth redistribution. And nationalizing essential public services. Now, thanks to all your boldness, we know that this isn't just a moral strategy. It can be a winning strategy. It fires up the base, and it activates constituencies that long ago stopped voting altogether.

You showed us something else in the last election, too, and it's just as important. You showed that political parties don't need to fear the creativity and independence of social movements—and social movements, likewise, have a huge amount to gain from engaging with electoral politics.

That's a very big deal. Because let's be honest: political parties tend to be a bit freakish about control. And real grassroots movements—we cherish our independence, and we're pretty much impossible to control. But what we are seeing with the remarkable relationship between Labour and Momentum†, and with other

*In the 2017 general elections, the Labour Party increased its share of votes by more than in any election since 1945. The Conservatives lost their majority but clung to power by forming a coalition with Ireland's Democratic Unionist Party (DUP).

†Momentum is a grassroots movement associated with the Labour Party that supports progressive candidates and pushes the party to the left.

wonderful campaign organizations, is that it is possible to combine the best of both worlds. If we listen and learn from one another, we can create a force that is both stronger and nimbler than anything either parties or movements can pull off on their own.

I want you to know that what you have done here is reverberating around the world. So many of us are watching your ongoing experiment in this new kind of politics with rapt attention. And, of course, what happened here is itself part of a global phenomenon. It's a wave led by young people who came into adulthood just as the global financial system was collapsing and just as climate disruption was banging down the door.

Many come out of social movements like Occupy Wall Street, and Spain's Indignados. They began by saying no—to austerity, to bank bailouts, to wars and police violence, to fracking and pipelines. But they came to understand that the biggest challenge is overcoming the way neoliberalism has waged war on our collective imagination, on our ability to truly believe in anything outside its bleak borders.

And so, these movements started to dream together, laying out bold and different visions of the future, and credible pathways out of crisis. And most important, they began engaging with political parties, to try to win power. We saw it in Bernie Sanders's historic campaign in the 2016 US Democratic primaries, which was powered by Millennials who know that safe centrist politics offers them no kind of safe future.

In these cases and others, electoral campaigns caught fire with stunning speed, faster than any genuinely transformative political program has in either Europe or North America in my lifetime. But still, in each case, not close enough. So, in this time between elections, it's worth thinking about how to make absolutely sure that next time, all our movements go all the way.

A big part of the answer is: Keeping it up. Keep building that "yes."

But take it even farther.

Outside the heat of a campaign, there is more time to deepen the relationships between issues and movements, so that our solutions address multiple crises at once. In all our countries, we can and must do more to connect the dots between economic injustice, racial injustice, and gender injustice. We need to understand and explain how all those ugly systems that place one group in a position of dominance over another (based on skin color, religious faith, gender, and sexual orientation) consistently serve the interests of power and money, and always have. They do it by keeping us divided, and keeping themselves protected.

And we have to do more to keep it front of mind that we are in a state of climate emergency, the roots of which are found in the same system of bottomless greed that underlies our economic emergency. But states of emergency, let's recall, can be catalysts for deep progressive victories.

So, let's draw out the connections between the gig economy, which treats human beings like a raw resource from which to extract wealth and then discard, and the dig economy, in which the extractive companies treat the earth with the very same disdain. And let's show exactly how we can move from that gig and dig economy to a society based on principles of care and repair; where the work of our caregivers and of our land and water protectors is respected and valued; a world where no one and nowhere is thrown away, whether in firetrap housing estates or on hurricane-ravaged islands.

I applaud the clear stand Labour has taken against fracking and for clean energy. Now we need to up our ambition and show exactly how battling climate change is a once-in-a-century chance to build a fairer and more democratic economy. Because as we rapidly transition off fossil fuels, we cannot replicate the wealth concentration and the injustices of the oil and coal economy, in which hundreds of billions in profits have been privatized and the tremendous risks are socialized.

We can and must design a system in which the polluters pay a very large share of the cost of transitioning off fossil fuels, and where we keep green energy in public and community hands. That way, revenues stay in your communities, to pay for child care and firefighters and other crucial services. And it's the only way to make sure that the green jobs that are created are union jobs that pay a living wage.

The motto needs to be "Leave the oil and gas in the ground, but leave no worker behind." And the best part? You don't need to wait until you get to Westminster to start this great transition. You can use the levers you have right now.

You can turn your Labour-controlled cities into beacons for the world transformed. A good start would be divesting your pensions from fossil fuels and investing that money in low-carbon social housing and green-energy cooperatives. That way people can begin to experience the benefits of the next economy before the next election, and know in their bones that, yes, there is, and always has been, an alternative.

In closing, I want to stress, as your international speaker at this convention, that none of this can be about turning any one nation into a progressive museum, or fortress. In wealthy countries like yours and mine, we need policies that reflect what we owe to the Global South, that own up to our role in destabilizing the economies and ecologies of poorer nations for a great many years.

For instance, during this epic hurricane season, we've heard a lot of talk of the "British Virgin Islands," the "Dutch Virgin

Islands," the "French Caribbean," and so on. Rarely was it seen as relevant to observe that these are not reflections of where Europeans like to holiday. They are reflections of the fact that so much of the vast wealth of empire was extracted from these islands as a direct result of human bondage—wealth that supercharged Europe's and North America's Industrial Revolution, setting us up as the super-polluters we are today. And that is intimately connected to the fact that the very future of these island nations is now at grave risk from the triple-threat of superstorms, sea level rise, and dying coral reefs.

What should this painful history mean to us today?

It means welcoming migrants and refugees. And it means paying our fair share to help many more countries ramp up justice-based green transitions of their own. Trump going rogue is no excuse to demand less of ourselves in the United Kingdom and Canada or anywhere else, for that matter. It means the opposite: that we have to demand more of ourselves, to pick up the slack until the United States manages to get its sewer system unclogged.

I firmly believe that all this work, challenging as it is, is a crucial part of the path to victory; that the more ambitious, consistent, and holistic you can be in painting a picture of the world transformed, the more credible a Labour government will become.

Because you went and showed us all that you can win. Now you *have* to win.

We all do.

Winning is a moral imperative. The stakes are too high, and time is too short, to settle for anything less.

CAPITALISM KILLED OUR CLIMATE MOMENTUM, NOT "HUMAN NATURE"

In the nick of time, a new political path to safety is presenting itself.

AUGUST 2018

ON SUNDAY, THE ENTIRE *NEW YORK TIMES MAGAZINE* WILL BE COMPOSED OF just one article on a single subject: the failure to confront the global climate crisis in the 1980s, a time when the science was settled and the politics seemed to align. Written by Nathaniel Rich, this work of history filled with insider revelations about roads not taken made me, on several occasions, swear out loud. And lest there be any question about the world-shattering magnitude of this failure, Rich's words are punctuated with full-page aerial photographs by George Steinmetz that wrenchingly document the rapid unraveling of planetary systems, from the rushing water where Greenland ice used to be to massive algae blooms in China's third-largest lake.

The novella-length piece represents the kind of media commitment that the climate crisis has long deserved but almost never received. We have all heard the various excuses for why the small matter of despoiling our only home just doesn't cut it as a compelling news story: "Climate change is too far off in the future"; "It's inappropriate to talk about politics when people are losing their lives to hurricanes and fires"; "Journalists follow the news, they don't make it—and politicians aren't talking about climate change"; and of course, "Every time we try, it's a ratings killer."

None of the excuses can mask the dereliction of duty. It has always been possible for major media outlets to decide, all on their own, that planetary destabilization is a huge news story, unquestionably the most consequential of our time. They always had the capacity to harness the skills of their reporters and photographers to connect abstract science to lived extreme weather events. And if they did so consistently, it would lessen the need for journalists to get ahead of politics because the more informed the public is about both the threat and the tangible solutions, the more they push their elected representatives to take bold action.

Which is why it was so exciting to see the *Times* throw the full force of its editorial machine behind Rich's opus—teasing it with a promotional video, kicking it off with a live event at the Times Center, and providing accompanying educational materials. That's also why it is so enraging that the piece is spectacularly wrong in its central thesis.

According to Rich, between the years 1979 and 1989, the basic science of climate change was understood and accepted, the partisan divide over the issue had yet to cleave, the fossil fuel companies hadn't started their misinformation campaign in earnest, and there was a great deal of global political momentum toward a bold and binding international emissions reduction agreement. Writing of the key period at the end of the 1980s, Rich says, "The conditions for success could not have been more favorable."

And yet we blew it—"we" being humans, who apparently are just too shortsighted to safeguard our own future. Just in case we missed the point of who and what is to blame for the fact that we are now "losing earth," Rich's answer is presented in a full-page callout: "All the facts were known, and nothing stood in our way. Nothing, that is, except ourselves."

Yep, you and me. Not, according to Rich, the fossil fuel companies who sat in on every major policy meeting described in the piece. (Imagine tobacco executives being repeatedly invited by the US government to come up with policies to ban smoking. When those meetings failed to yield anything substantive, would we conclude that the reason was that humans just want to die? Might we perhaps determine instead that the political system is corrupt and busted?)

Many climate scientists and historians have pointed out this misreading since the piece went online.* Others have remarked on the maddening invocations of "human nature" and the use of the royal "we" to describe a screamingly homogenous group of US power players. Throughout Rich's accounting, we hear nothing from those political leaders in the Global South who were demanding binding action in this key period and after, somehow able to care about future generations despite being human. The voices of women, meanwhile, are almost as rare in Rich's text as sightings of the endangered ivory-billed woodpecker, and when we ladies do appear, it is mainly as long-suffering wives of tragically heroic men.

All these flaws have been well covered, so I won't rehash them here. My focus is the central premise of the piece: that the end of the 1980s presented conditions that "could not have been more favorable" to bold climate action. On the contrary, one could

*When Rich expanded the article into a book in 2019, he corrected the omission.

scarcely imagine a more inopportune moment in human evolution for our species to come face-to-face with the hard truth that the conveniences of modern consumer capitalism were steadily eroding the habitability of the planet. Why? Because the late '80s was the absolute zenith of the neoliberal crusade, a moment of peak ideological ascendency for the economic and social project that deliberately set out to vilify collective action in the name of liberating "free markets" in every aspect of life. Yet Rich makes no mention of this parallel upheaval in economic and political thought.

When I delved into this same climate change history some years ago, I concluded, as Rich does, that the key juncture when world momentum was building toward a tough, science-based global agreement was 1988. That was when James Hansen, then director of NASA's Goddard Institute for Space Studies, testified before Congress that he had "99 percent confidence" in "a real warming trend" linked to human activity. Later that same month, hundreds of scientists and policymakers held the historic World Conference on the Changing Atmosphere in Toronto, where the first emission-reduction targets were discussed. By the end of that same year, in November 1988, the UN Intergovernmental Panel on Climate Change, the premier scientific body advising governments on the climate threat, held its first session.

But climate change wasn't a concern just for politicians and wonks back then—it was watercooler stuff, so much so that when the editors of *Time* magazine announced their 1988 "Man of the Year," they went for "Planet of the Year: Endangered Earth." The cover featured an image of the globe held together with twine, the sun setting ominously in the background. "No single individual, no event, no movement captured imaginations or dominated headlines more," journalist Thomas Sancton explained, "than the clump of rock and soil and water and air that is our common home."

(Interestingly, unlike Rich, Sancton didn't blame "human nature" for the planetary mugging. He went deeper, tracing it to the misuse of the Judeo-Christian concept of "dominion" over nature and the fact that it supplanted the pre-Christian idea that "the earth was seen as a mother, a fertile giver of life. Nature—the soil, forest, sea—was endowed with divinity, and mortals were subordinate to it.")

When I surveyed the climate news from this period, it really did seem like a profound shift was within grasp—and then, tragically, it all slipped away, with the United States walking out of international negotiations and the rest of the world settling for nonbinding agreements that relied on dodgy "market mechanisms" like carbon trading and offsets and, in a few rare cases, a minor carbon tax. So, it really is worth asking, as Rich does: What the hell happened? What interrupted the urgency and determination that were emanating from all these elite institutions simultaneously by the end of the '80s?

Rich concludes, while offering no social or scientific evidence, that something called "human nature" kicked in and messed everything up. "Human beings," he writes, "whether in global organizations, democracies, industries, political parties or as individuals, are incapable of sacrificing present convenience to forestall a penalty imposed on future generations." It seems we are wired to "obsess over the present, worry about the medium term and cast the long term out of our minds, as we might spit out a poison."

When I looked at the same period, I came to a very different conclusion: that what at first seemed like our best shot at life-saving climate action had in retrospect suffered from an epic case of historical bad timing. Because what becomes clear when you look back at this juncture is that just as governments were getting together to get serious about reining in the fossil fuel sector, the global neoliberal revolution went supernova, and that project of economic and social reengineering clashed with the imperatives of both climate science and corporate regulation at every turn.

The failure to make even a passing reference to this other global trend that was unfolding in the late '80s represents an unfathomably large blind spot in Rich's piece. After all, the primary benefit of returning to a period in the not-too-distant past as a journalist is that you are able to see trends and patterns that were not yet visible to people living through those tumultuous events in real time. The climate community in 1988, for instance, had no way of knowing that they were on the cusp of the convulsive economic revolution that would remake every major economy on the planet.

But we know. And one thing that becomes very clear when you look back on the late '80s is that far from offering "conditions for success [that] could not have been more favorable," 1988–89 was *the worst possible moment* for humanity to decide that it was going to get serious about putting planetary health ahead of profits.

Recall what else was going on. In 1988, Canada and the United States signed their Free Trade Agreement, a prototype for NAFTA and countless deals that would follow. The Berlin Wall was about to fall, an event that would be successfully seized upon by right-wing ideologues in the United States as proof of "the end of history" and taken as license to export the Reagan-Thatcher recipe for privatization, deregulation, and economic austerity to every corner of the globe.

It was this convergence of historical trends—the emergence of a global architecture that was supposed to tackle climate change *and* the emergence of a much more powerful global architecture to liberate capital from all constraints—that derailed the momentum Rich rightly identifies. Because, as he notes repeatedly, meeting the challenge of climate change would have required imposing stiff regulations on polluters while investing in the public sphere to transform how we power our lives, live in cities, and move ourselves around.

All this was possible in the 1980s and '90s—it still is today—but it would have demanded a head-on battle with the project of neoliberalism, which at that time was waging war on the very idea of the public sphere. ("There is no such thing as society," Thatcher told us.) Meanwhile, the free trade deals being signed in this period were busily making many sensible climate initiatives (like subsidizing and offering preferential treatment to local green industry and refusing many polluting projects like fracking and oil pipelines) illegal under international trade law.

As I wrote in *This Changes Everything*, "We have not done the things that are necessary to lower emissions because those things fundamentally conflict with deregulated capitalism, the reigning ideology for the entire period that we have been struggling to find a way out of this crisis. We are stuck because the actions that would give us the best chance of averting catastrophe, and that would benefit the vast majority, are extremely threatening to an elite minority that has a stranglehold over our economy, our political process, and most of our major media outlets."

Why does it matter that Rich makes no mention of this clash and, instead, claims our fate has been sealed by "human nature"? It matters because if the force that interrupted the momentum toward action is "ourselves," then the fatalistic headline LOSING EARTH really is merited. If an inability to sacrifice in the short term for a shot at health and safety in the near future is baked into our collective DNA, then we have no hope of turning things around in time to avert truly catastrophic warming.

If, on the other hand, we humans really were on the brink of saving ourselves in the '80s, but were swamped by a tide of elite, free-market fanaticism, one that was opposed by millions of people around the world, then there is something quite concrete we can do about it. We can confront that economic order and try to replace it with something that is rooted in both human and planetary security, one that does not place at its center the quest for growth and profit at all costs.

And the good news—and, yes, there is some—is that today, unlike in 1989, a young and growing movement of green Democratic Socialists is advancing in the United States with precisely that vision. And that represents more than just an electoral alternative—it's our one and only planetary lifeline.

Yet we have to be clear that the lifeline we need is not something that has been tried before, at least not at anything like the scale required. When the *Times* tweeted out its teaser for Rich's article about "humankind's inability to address the climate change catastrophe," the excellent eco-justice wing of the Democratic Socialists of America quickly offered this correction: "*CAPITALISM* If they were serious about investigating what's gone so wrong, this would be about 'capitalism's inability to address the climate change catastrophe.' Beyond capitalism, *humankind* is fully capable of organizing societies to thrive within ecological limits."

Their point is a good one, if incomplete. There is nothing essential about humans living under capitalism; we humans are capable of organizing ourselves into all kinds of different social orders, including societies with much longer time horizons and far more respect for natural life-support systems. Indeed, humans have lived that way for the vast majority of our history, and many Indigenous cultures keep Earth-centered cosmologies alive to this day. Capitalism is a tiny blip in the collective story of our species.

But simply blaming capitalism isn't enough. It is absolutely true that the drive for endless growth and profits stands squarely opposed to the imperative for a rapid transition off fossil fuels. It is absolutely true that the global unleashing of the unbound form of capitalism known as neoliberalism in the '80s and '90s has been the single greatest contributor to a disastrous global emission spike in recent decades, and the single greatest obstacle to science-based climate action since governments began meeting to talk (and talk and talk) about lowering emissions. And it remains the biggest obstacle today, even in countries that market themselves as climate leaders.

But we have to be honest that autocratic industrial socialism has also been a disaster for the environment, as evidenced most dramatically by the fact that carbon emissions briefly plummeted when the economies of the former Soviet Union collapsed in the early 1990s. And Venezuela's petro-populism is a reminder that there is nothing inherently green about self-defined socialism.

Let's acknowledge this fact, while also pointing out that countries with strong democratic-socialist tradition (like Denmark, Sweden, and Uruguay) have some of the most visionary environmental policies in the world. From this we can conclude that socialism isn't necessarily ecological, but that a new form of democratic eco-socialism, with the humility to learn from Indigenous teachings about the duties to future generations and the interconnection of all life, appears to be humanity's best shot at collective survival.

These are the stakes in the surge of movement-grounded politicians and political candidates who are advancing a democratic eco-socialist vision, connecting the dots between the economic depredations caused by decades of neoliberal ascendency and the ravaged state of our natural world. Together they are calling for a Green New Deal that meets everyone's basic material needs and offers real solutions to racial and gender inequities, all while catalyzing a rapid transition to 100 percent renewable energy. Many have also pledged not to take money from fossil fuel companies and are promising instead to prosecute them.

This new generation of political leaders is rejecting the neoliberal centrism of the Democratic Party establishment, with its tepid "market-based solutions" to the ecological crisis, as well as Donald Trump's all-out war on nature. And they are also presenting a concrete alternative to the extractivist socialists of both the past and present. Perhaps most important, this new generation of leaders isn't interested in scapegoating "humanity" for the greed and corruption of a tiny elite. It seeks instead to help humanity, particularly its most systematically unheard and uncounted members, to find their collective voice and power so they can stand up to that elite.

We aren't losing the earth, but the earth is getting so hot so fast that it is on a trajectory to lose a great many of us. In the nick of time, a new political path to safety is presenting itself. This is no moment to bemoan our lost decades. It's the moment to get the hell on that path.

THERE'S NOTHING NATURAL ABOUT PUERTO RICO'S DISASTER

When you systematically starve and neglect the very bones of a society, rendering it dysfunctional on a good day, that society has absolutely no capacity to weather a true crisis.

SEPTEMBER 2018, ONE YEAR AFTER HURRICANE MARIA

FOR A COUPLE OF DECADES, I'VE BEEN INVESTIGATING THE WAYS THAT THE already rich and powerful systematically exploit the pain and the trauma of collective shocks (like superstorms or economic crises) in order to build an even more unequal and undemocratic society.

Long before Hurricane Maria, Puerto Rico was a textbook example. Before those fierce winds came, the debt (illegitimate and much of it illegal) was the excuse used to ram through a brutal program of economic suffering, what the great Argentine author Rodolfo Walsh, writing about four decades earlier, famously called *miseria planificada*, "planned misery."

This program systematically attacked the very glue that holds a society together: all levels of education, health care, the electricity and water systems, transit systems, communication networks, and more.

It was a plan so widely rejected that no elected representatives in Puerto Rico could be trusted to carry it out—which is why in 2016 the US Congress passed the Puerto Rico Oversight, Management, and Economic Stability Act, known as PROMESA. That law amounted to a financial coup d'état that put the territory's economy directly in the hands of the unelected Financial Oversight and Management Board. In Puerto Rico, they call it La Junta.

The term fits. As Greece's former finance minister Yanis Varoufakis puts it, governments used to be overthrown with tanks—"now it's with banks."

It was in this context, with every Puerto Rican institution already trembling from La Junta's assaults, that Maria's ferocious winds came roaring through. It was a storm so powerful it would have sent even the sturdiest society reeling. But Puerto Rico didn't just reel. Puerto Rico broke.

Not the people of Puerto Rico, but all those systems that had already been deliberately brought to the brink: power, health, water, communication, food. All those systems collapsed. The latest research puts the numbers of lives lost as a result of Hurricane Maria at approximately three thousand, a figure now accepted by the governor of Puerto Rico. But let us be clear: Maria didn't kill all those people. It was that *combination* of grinding austerity and an extraordinary hurricane that stole so many precious lives.

A few lives were lost to wind and water, yes. But the vast majority died because when you systematically starve and neglect the very bones of a society, rendering it dysfunctional on a good day, that society has absolutely no capacity to weather a true crisis. That is what the research tells us, those studies Donald Trump so casually denies: The major causes of death were people being unable to plug in medical equipment because the electricity grid was down for months; health networks so diminished they were unable to provide medicine for treatable diseases. People died because they were left to drink contaminated water because of a legacy of environmental racism. People died because they were abandoned and left without hope for so long that suicide seemed the only option.

Those deaths were not the result of an unprecedented "natural disaster" or even "an act of God," as we so often hear.

Honoring the dead begins with telling the truth. And the truth is that there is nothing natural about this disaster. And if you believe in God, leave her out of this, too.

God isn't the one who laid off thousands of skilled electrical workers in the years before the storm, or who failed to maintain the grid with basic repairs. God didn't give vital relief and reconstruction contracts to politically connected firms, some of whom didn't even pretend to do their jobs. God didn't decide that Puerto Rico should import 85 percent of its food—this archipelago blessed with some of the most fertile soil in the world. God didn't decide Puerto Rico should get 98 percent of its energy from imported fossil fuels—these islands bathed in sun, lashed by wind, and surrounded by waves, all of which could provide cheap and clean renewable power to spare.

These were decisions made by people working for powerful interests.

Because for five hundred uninterrupted years, the role of Puerto Rico and Puerto Ricans in the world economy has been to make other people rich, whether with the extraction of cheap labor or cheap resources or by being a captive market for imported food and fuel.

A colonial economy by definition is a dependent economy; a

centralized, lopsided, and distorted economy. And as we have seen, an intensely vulnerable economy.

And it isn't even right to call the storm itself a "natural disaster." None of these record-breaking storms are natural anymore—Irma and Maria, Katrina and Sandy, Haiyan and Harvey, and now Florence and Super Typhoon Mangkhut. The reason we are seeing records shattered time after time is that the oceans are warmer and the tides are higher. And that's not God's fault, either.

This is the deadly cocktail—not just a storm, but a storm supercharged by climate change slamming headlong into a society deliberately weakened by a decade of unrelenting austerity layered on top of centuries of colonial extraction, with relief efforts that make no attempt to disguise the fact that the lives of the poor exist within our global system at a sharp discount.

Maria blew so hard that she tore all the genteel disguises off these brutal systems just as she blew the leaves off the trees, leaving them naked for the world to see. The hurricane and FEMA's endless failures pushed Puerto Rico over the edge. But we have to face up to why the territory was teetering so precariously on the precipice in the first place.

We also need to stop framing these failures as incompetence. Because if it were incompetence, there would be some effort to fix the underlying systems that produced the failures; to rebuild the public sphere, design a more secure food and energy system, and stop the carbon pollution that guarantees even more ferocious storms in the coming decades.

Yet we have seen the precise opposite. We have seen nothing but more disaster capitalism using the trauma of the storm to push massive cuts to education, hundreds of school closures, wave after wave of home foreclosures, and the privatization of some of Puerto Rico's most valuable assets. And just as Trump denies the reality of thousands of Puerto Rican deaths, he also denies the reality of climate change—which his administration must do in order to push dozens of toxic policies that make the crisis even worse.

Such is the official response to this modern-day catastrophe: Do everything possible to make sure that it will happen again and again. Do everything possible to bring on a future in which climate disasters arrive so fast and so furious that even gathering together to mourn the dead on painful anniversaries could, for our children, come to seem like an unattainable luxury. They will already be in the throes of the next emergency, like people in North and South Carolina, in South India, and in the Philippines are right now, exactly one year after Maria made landfall.

That is why dozens of Puerto Rican organizations, under the banner of *Junte Gente*, "the People Together," are standing up to demand a different future. Not just a little bit better but radically better. Their message is a clear one: that this storm must be a wakeup call, a historic catalyst for a just recovery and a just transition to the next economy. Right now.

That begins with auditing and ultimately erasing the island's illegal debt and firing La Junta, whose very existence is an affront to the most basic principles of self-government. Only then will there be political space to redesign the food, energy, housing, and transportation systems that failed so many and replace them with institutions that truly serve the Puerto Rican people.

This movement for a just recovery draws on local brilliance and protected knowledge to make the most of the richness of the soil to feed the people and on the power of the sun and wind to provide energy to the archipelago.

Today I am reminded of the words of Dalma Cartagena, one of the great leaders of Puerto Rico's agro-ecology movement, which has been spearheading the calls for the island to stop relying on imported food and to build resilience by reviving traditional agricultural practices. "Maria hit us hard," she said. "But it made our convictions stronger. Made us know the correct path."

The era of planned misery and deliberately designed dependence is ending. It's time to plan for joy and design for liberation, so that when the next storm comes—and it will—the winds will roar and the trees will bend, but Puerto Rico will show the world that it can never be broken.

MOVEMENTS WILL MAKE, OR BREAK, THE GREEN NEW DEAL

We have been trained to see our issues in silos; they never belonged there.

FEBRUARY 2019

"I REALLY DON'T LIKE THEIR POLICIES OF TAKING AWAY YOUR CAR, TAKING AWAY your airplane flights, of 'let's hop a train to California,' or 'you're not allowed to own cows anymore!'"

So bellowed President Donald Trump in El Paso, Texas, in his first campaign-style salvo against Representative Alexandria Ocasio-Cortez and Senator Ed Markey's Green New Deal resolution.

It's worth marking the moment. Because those could be the famous last words of a one-term president who wildly underestimated the public appetite for transformative action on the triple crises of our time: imminent ecological unraveling, gaping economic inequality (including the racial and gender wealth gaps), and surging white supremacy.

Or they could be the epitaph for a habitable climate, with Trump's lies and scare tactics succeeding in trampling this desperately needed framework; that could either help win him reelection or land us with a timid Democrat in the White House with neither the courage nor the democratic mandate for this kind of deep change. Either scenario means blowing the handful of years left to roll out the transformations required to keep temperatures below catastrophic levels.

In October 2018, the Intergovernmental Panel on Climate Change published its landmark report informing us that global emissions need to be slashed in half in less than twelve years, a target that simply cannot be met without the world's largest economy playing a game-changing leadership role. If there is a new administration ready to leap into that role in January 2021, meeting those targets will still be extraordinarily difficult, but it will be technically possible—especially if large cities and states like California and New York continue to escalate their ambitions in the interim, along with the European Union, which is in the midst of its own Green New Deal debate. Losing another four years to a Republican or a corporate Democrat, and starting in 2026 is, quite simply, a joke.

So, either Trump is right and the Green New Deal is a losing political issue, one he can smear out of existence, or he is wrong and a candidate who makes the Green New Deal the centerpiece of their platform will take the Democratic primary and then defeat Trump in the general, with a clear democratic mandate to introduce wartime levels of investment to battle our triple crises from day one. That would very likely inspire the rest of the world to finally follow suit on bold climate policy, giving us all a fighting chance.

The good news is that, as I write, there are candidates vying for the leadership of the Democratic Party (most notably Bernie Sanders and Elizabeth Warren) who have not only endorsed the Green New Deal, but who also have a proven track record of standing up to the two most powerful industries trying to block it: fossil fuel companies and the banks that finance them. These leaders (and the movements that made them) understand something critical about the transition we need: It won't all be win-win. For any of this to happen, fossil fuel companies, which have made obscene profits for many decades, will have to start losing, and losing more than just the tax breaks and subsidies to which they are so accustomed. They will also have to lose the new drilling and mining leases they want; they'll have to be denied permits for the pipelines and export terminals they very much want to build. They will have to leave trillions of dollars' worth of proven fossil fuel reserves in the ground. They may even have to divert their remaining profits to paying for the mess they knowingly made, as several lawsuits are attempting to establish.

Meanwhile, if we have smart policies in place to encourage solar panels to proliferate on rooftops, big power utilities will lose a significant portion of their profits, since their former customers will be in the energy-generation business. This would create huge opportunities for a more level economy and, ultimately, for lower utility bills—but once again, some powerful interests will have to lose, namely the huge coal-powered utilities who have no interest in watching as their onetime captive customers turn into competitors, selling power back to the grid.

Politicians willing to inflict these losses on fossil fuel companies and their allies need to be more than just not actively corrupt. They need to be up for the fight of the century—and absolutely clear about which side must win. But even then, there is one more element we must never forget: any administration attempting to implement a Green New Deal will need powerful social movements both backing them up and pushing them to do more.

Indeed, the single largest determining factor in whether a Green New Deal mobilization pulls us back from the climate cliff will be the actions taken by social movements in the coming years. Because as important as it is to elect politicians who are up for this fight, the decisive questions are not going to be settled through elections alone. At their core, they are about building political power—enough to change the calculus of what is possible.

This is the overarching lesson from those few-and-far-between chapters in history when the governments of wealthy countries agreed to introduce big changes to the building blocks of their economies. It must always be remembered that President Franklin D. Roosevelt rolled out the New Deal in the midst of a historic wave of labor unrest: There was the Teamster Rebellion and the Minneapolis general strike in 1934, the eighty-three-day shutdown of West Coast ports by longshore workers that same year, and the Flint autoworkers sit-down strikes in 1936 and 1937.

During this same period, mass movements, responding to the suffering of the Great Depression, demanded sweeping social programs, such as Social Security and unemployment insurance, while socialists argued that abandoned factories should be handed over to their workers and turned into cooperatives. Upton Sinclair, the muckraking author of *The Jungle*, ran for governor of California in 1934 on a platform arguing that the key to ending poverty was full state funding of workers' cooperatives. He received nearly 900,000 votes, but having been viciously attacked by the right and undercut by the Democratic establishment, he fell just short of winning the governor's office. Growing numbers of Americans were also paying close attention to Huey Long, the populist senator from Louisiana who believed that all Americans should receive a guaranteed annual income of $2,500. Explaining why he had added more social welfare benefits to the New Deal in 1935, FDR said he wanted to "steal Long's thunder."

All this is a reminder that the New Deal was adopted by Roosevelt at a time of such progressive and left militancy that its programs (which seem radical by today's standards) appeared at the time to be the only way to hold back a full-scale revolution.

A similar dynamic was at play in 1948, when the United States decided to underwrite the Marshall Plan. With Europe's infrastructure shattered and its economies in crisis, the US government was worried that large parts of Western Europe would see the egalitarian promises of socialism as their best hope and fall under the influence of the Soviet Union. Indeed, so many Germans were drawn to socialism after the war that the Allied Powers decided to split Germany into two parts rather than risk losing it all to the Soviets.

It was in this context that the US government decided it would not rebuild Western Germany with Wild West capitalism (as it would attempt to do five decades later when the Soviet Union collapsed, with disastrous results). Rather, Germany would be rebuilt on a mixed social-democratic model, with supports for local industry, strong trade unions, and a robust welfare state. As with the New Deal, the idea was to build a market economy with enough socialist elements that a more revolutionary approach would be drained of its appeal. Carolyn Eisenberg, author of an acclaimed history of the Marshall Plan, stresses that this approach was not born of altruism. "The Soviet Union was like a loaded gun. The economy was in crisis, there was a substantial German left, and they [the West] had to win the allegiance of the German people fast."

This pressure from the left, in the form of militant movements and political parties, delivered the most progressive elements of the New Deal and the Marshall Plan. That's important to remember because the Green New Deal plans currently on offer from political parties in North America and Europe still have significant weaknesses and will need to be toughened and expanded, just as the original New Deal was over time.

The Ocasio-Cortez and Markey resolution is a loose framework, and as much as it has been criticized in the press for including too much, the reality is that it still leaves a lot out. For instance, a

Green New Deal needs to be more explicit about keeping carbon in the ground, about the central role of the US military in driving up emissions, about nuclear and coal never being "clean," and about the debts wealthy countries like the United States and powerful corporations like Shell and Exxon owe to poorer nations that are coping with the impacts of crises they did almost nothing to create.

Most fundamentally, any credible Green New Deal needs a concrete plan for ensuring that the salaries from all the good green jobs it creates aren't immediately poured into high-consumer lifestyles that inadvertently end up increasing emissions—a scenario where everyone has a good job and lots of disposable income and it all gets spent on throwaway crap imported from China destined for the landfill.

This is the problem with what we might call the emerging "climate Keynesianism": the post–World War II economic boom did revive ailing economies, but it also kicked off suburban sprawl and set off a consumption tidal wave that would eventually be exported to every corner of the globe. In truth, policymakers are still dancing around the question of whether we are talking about slapping solar panels on the roof of Walmart and calling it green, or whether we are ready to have a more probing conversation about the limits of lifestyles that treat shopping as the main way to form identity, community, and culture.

That conversation is intimately connected to the kinds of investments we prioritize in our Green New Deals. What we need are transitions that recognize the hard limits on extraction and that simultaneously create new opportunities for people to improve quality of life and derive pleasure outside the endless consumption cycle, whether through publicly funded art and urban recreation or access to nature through new protections for wilderness. Crucially, that means making sure that shorter work weeks allow people the time for this kind of enjoyment, and that they are not trapped in the grind of overwork requiring the quick fixes of fast food and mind-numbing distraction.

We already know that these are the kinds of lifestyle changes and leisure activities that tangibly increase happiness and fulfillment but, particularly in the US, debates about climate action remain trapped in a paradigm that equates quality of life with personal prosperity and wealth accumulation. If the political roadblocks to a Green New Deal are to be broken, this equation will need to be broken too.

As the *Guardian*'s George Monbiot puts it, our planet's resources can provide us with "private sufficiency and public luxury," in the forms of "wonderful parks and playgrounds, public sports centres and swimming pools, galleries, allotments and public transport networks." The earth cannot, however, sustain the impossible dream of private luxury for all. This is what economist Kate Raworth calls for in her book *Doughnut Economics*: "meeting the needs of all within the means of the planet" through economies that "make us thrive, whether or not they grow."

In this regard, there is much to learn from Indigenous-led movements in Bolivia and Ecuador that have placed at the center of their calls for ecological transformation the concept of *buen vivir*, a focus on the right to a good life as opposed to the more-and-more life of ever escalating consumption and planned obsolescence.

Opponents of the Green New Deal can be counted on to continue spreading fear that what is being proposed is an austere future marked by nonstop deprivation and government controls. The response cannot be to deny that there will be changes to the way the wealthiest 10–20 percent of humanity has come to live. There will be changes, there will be areas where we in this category must contract—including air travel, meat consumption, and profligate energy use—but there will also be new pleasures and new spaces where we can build abundance.

As we have these difficult debates, we also need to remember that the health of our planet is the single greatest determining factor in the quality of all our lives. And having waded through more than my share of wreckage after hurricanes and superstorms, from Katrina to Sandy to Maria, and inhaled too much air choked with the particulates from too many spontaneously combusting forests, I feel confident in saying that a climate-disrupted future is a bleak and an austere future, one capable of turning all our material possessions into rubble or ash with terrifying speed. We can pretend that extending the status quo into the future, unchanged, is one of the options available to us. But that is a fantasy. Change is coming one way or another. Our choice is whether we try to shape that change to the maximum benefit of all or wait passively as the forces of climate disaster, scarcity, and fear of the "other" fundamentally reshape us.

All this is why there must be rigorous checks and balances—including regular carbon audits—built in to every country's Green New Deal to make sure that we actually hit the steep emission-reduction targets mandated by science. If we simply assume that by switching to renewables and building energy-efficient housing it will happen on its own, we could end up in the supremely ironic situation of kicking off a Green New Deal emissions spike.

In short, the Green New Deal will necessarily be a work in progress, one that is only as robust as the social movements, unions, scientists, and local communities that are pushing for it to live up to its promise. Right now, civil society is nowhere near as strong or as organized as it was in the 1930s, when the huge concessions of the New Deal era were won—though there are certainly signs of strength, from movements against mass incarceration and deportations, to #MeToo, to the wave of teachers' strikes, to Indigenous-led pipeline blockades, to fossil fuel divestment, to the Women's

Marches, to School Strikes for Climate, to the Sunrise Movement, to the momentum for Medicare for All, and much more.

Still, there remains a long way to go to build the kind of outside power required to win and protect a truly transformational Green New Deal, which is why it is so crucial that we use the existing framework as a potent tool to build that power—a vision to both unite movements that are not currently in conversation with one another and to dramatically expand all their bases.

Central to that project is turning what is being derided as a left-wing "laundry list" or "wish list" into an irresistible story of the future, connecting the dots among the many parts of daily life that stand to be transformed, from health care to employment, daycare to jail cell, clean air to leisure time.

Right now, the Green New Deal is being characterized as an unrelated grab bag because most of us have been trained to avoid a systemic and historical analysis of capitalism and to divide pretty much every crisis our system produces (economic inequality, violence against women, white supremacy, unending wars, ecological unraveling) into walled-off silos. From within that rigid mind-set, it's easy to dismiss a sweeping and intersectional vision like the Green New Deal as a green-tinted "laundry list" of everything the left has ever wanted.

For this reason, one of the most pressing tasks ahead is to use every tool possible to make the case for how our overlapping crises are indeed inextricably linked—and can be overcome only with a holistic vision for social and economic transformation. We can point out, for instance, that no matter how fast we move to lower emissions, it is going to get hotter and storms are going to get fiercer. When those storms bash up against health care systems that have been starved by decades of austerity, thousands pay the price with their lives, as they so tragically did in post-Maria Puerto

Rico. That's why putting universal health care in the Green New Deal is not an opportunistic add-on—it's an essential part of how we will keep our humanity in the stormy future ahead.

And there are many more connections to be drawn. Those complaining about climate policy being weighed down by supposedly unrelated demands for child care and free postsecondary education would do well to remember that the caring professions (most of them dominated by women) are relatively low carbon and can be made even more so with smart planning. In other words, they deserve to be seen as "green jobs," with the same protections, the same investments, and the same living wages as male-dominated workforces in the renewables, efficiency, and public transit sectors. Meanwhile, to make those sectors less male-dominated, family leave and pay equity are a must, which is why both are included in the Green New Deal resolution. We have been trained to see our issues in silos; they never belonged there.

Drawing out these connections in ways that capture the public imagination will take a massive exercise in participatory democracy. A first step is for workers in every sector (hospitals, schools, universities, tech, manufacturing, media, and more) to make their own plans for how to rapidly decarbonize while furthering the Green New Deal's mission to eliminate poverty, create good jobs, and close the racial and gender wealth divides. The Green New Deal resolution explicitly calls for this kind of democratic, decentralized leadership, and making it happen would go a long way toward building the broad base of support this framework will need to take on the powerful elite forces that are already lining up against it.

And there are plenty more connections to be made. A job guarantee, far from an unrelated socialist addendum, is a critical part of achieving a rapid and just transition. It would immediately lower the intense pressure on workers to take the kinds of jobs that destabilize our planet because all would be free to take the time needed to retrain and find work in one of the many sectors that will be dramatically expanding.

All these so-called bread-and-butter provisions (for job security, health care, child care, education, and housing) are fundamentally about creating a context in which the rampant economic insecurity of our age is addressed at the source. And that has everything to do with our capacity to cope with climate disruption, because the more secure people feel, knowing that their families will not want for food, medicine, and shelter, the less vulnerable they will be to the forces of racist demagoguery that will prey on the fears that invariably accompany times of great change. Put another way, this is how we are going to address the crisis of empathy in a warming world.

One last connection I will mention has to do with the concept of "repair." The resolution calls for creating well-paying jobs, "restoring and protecting threatened, endangered, and fragile ecosystems," and "cleaning up existing hazardous waste and abandoned sites, ensuring economic development and sustainability on those sites."

There are many such sites across the United States, entire landscapes that have been left to rot after they were no longer useful to frackers, miners, and drillers. It's a lot like how this culture treats people. It's certainly how we have been trained to treat our stuff—use it once, or until it breaks, then throw it away and buy some more. It's similar to what has been done to so many workers in the neoliberal period: they are used up and then abandoned to addiction and despair. It's what the entire carceral state is about: locking up huge sectors of the population who are more economically valuable as prison laborers and numbers on the spreadsheet of a private prison than they are as free workers.

There is a grand story to be told here about the duty to repair—to repair our relationship with the earth and with one another.

Because while it is true that climate change is a crisis produced by an excess of greenhouse gases in the atmosphere, it is also, in a more profound sense, a crisis produced by an extractive mind-set, by a way of viewing both the natural world and the majority of its inhabitants as resources to use up and then discard. I call it the "gig and dig" economy and firmly believe that we will not emerge from this crisis without a shift in worldview at every level, a transformation to an ethos of care and repair. Repairing the land. Repairing our stuff. Fearlessly repairing our relationships within our countries and between them.

We must always remember that the fossil fuel era began in violent kleptocracy, with those two foundational thefts of stolen people and stolen land that kick-started a new age of seemingly endless expansion. The route to renewal runs through reckoning and repair: reckoning with our past and repairing relationships with the people who paid the steepest price of the first Industrial Revolution.

These failures to confront difficult truths have long made a mockery of any notion of a collective "we"; only when we reckon with them will our societies be liberated to find our collective purpose. In fact, delivering that sense of common purpose is perhaps the Green New Deal's greatest promise. Because it isn't only the planet's life support systems that are unraveling before our eyes. So too is our social fabric, on so many fronts at once.

The signs of fracture are all around—from the rise of fake news and unhinged conspiracy theories to the hardened arteries of our body politic. In this context, a Green New Deal, precisely because of its sweeping scale, ambition, and urgency, could be the collective purpose that finally helps overcome many of these divides.

It's not a magic cure for racism or misogyny or homophobia or transphobia—we still have to confront those evils head on. But if it became law, despite all the powers arrayed against it, it would give a great many of us a sense of working together toward something bigger than ourselves. Something we are all a part of creating. And it would give us a shared destination—somewhere distinctly better than where we are now. That kind of shared mission is something our late capitalist culture badly needs right now.

If these kinds of deeper connections between fractured people and a fast-warming planet seem far beyond the scope of policy-makers, it's worth thinking back to the absolutely central role of artists during the New Deal era. Playwrights, photographers, muralists, and novelists were all part of telling the story of what was possible. For the Green New Deal to succeed, we, too, will need the skills and expertise of many different kinds of storytellers: artists, psychologists, faith leaders, historians, and more.

The Green New Deal framework has a way to go before everyone sees their future in it. Mistakes have already been made, and more will be made along the way. But none of this is as important as what this fast-growing political project gets exactly right.

The Green New Deal will need to be subject to constant vigilance and pressure from experts who understand exactly what it will take to lower our emissions as rapidly as science demands, and from social movements that have decades of experience bearing the brunt of pollution and false climate solutions. But in remaining vigilant, we also have to be careful not to lose sight of the big picture: that this is a potential lifeline that we all have a sacred and moral responsibility to reach for.

The young organizers in the Sunrise Movement, who have done so much to galvanize the Green New Deal momentum, talk about our collective moment as one filled with both "promise and peril." That is exactly right. And everything that happens from here on should hold one in each hand.

THE ART OF THE GREEN NEW DEAL

"We didn't just change the infrastructure. We changed how we did things. We became a society that was not only modern and wealthy, but dignified and humane."

APRIL 2019

SOMETIMES A PROJECT TAPS INTO A FORCE THAT IS POWERFUL WELL BEYOND the expectations of its creators. So it was with *A Message from the Future with Alexandria Ocasio-Cortez*, a seven-minute video I executive-produced and conceived of with the artist Molly Crabapple.

Narrated by the congresswoman and illustrated by Crabapple, the film is set a couple of decades from now. It begins with Ocasio-Cortez, a white streak in her hair, riding the bullet train from New York to Washington, DC. Rushing past the window is the future created by the successful implementation of a Green New Deal.

The film project grew out of a conversation I had with Crabapple (a brilliant illustrator, writer, and filmmaker) shortly after the idea for a Green New Deal started gaining traction in the United

States. We were brainstorming about how to involve more artists in the project. Most art forms are pretty low carbon, after all, and Franklin D. Roosevelt's New Deal led to a renaissance of publicly funded art, with artists of every stripe directly participating in the era's transformations.

We wanted to try to galvanize artists into that kind of social mission again, but not years down the road, if the Green New Deal became federal law. No, we wanted to see art right away, to help win the battle for hearts and minds that would determine whether the Green New Deal had a fighting chance in the first place.

Crabapple suggested doing a film on the Green New Deal with Ocasio-Cortez as the narrator and herself as illustrator. The question was: How do we tell the story of something that hasn't happened yet?

As we threw ideas around, we realized that your standard "explainer" video wouldn't cut it. The biggest obstacle to the kind of transformative change that the Green New Deal envisions is not that people fail to understand what is being proposed (though there is certainly plenty of misinformation floating around). It's that so many are convinced that humanity could never pull off something at this scale and speed. And a whole lot of people have come to believe that dystopia is a foregone conclusion.

The skepticism is understandable. The idea that societies could collectively decide to embrace rapid foundational changes to transportation, housing, energy, agriculture, forestry, and more—precisely what is needed to avert climate breakdown—is not something for which most of us have any living reference. We have grown up bombarded with the message that there is no alternative to the crappy system that is destabilizing the planet and hoarding vast wealth at the top. From most economists, we hear that we are fundamentally selfish, gratification-seeking units. From historians, we learn that social change has always been the work of singular great men.

Hollywood hasn't been much help, either. Almost every vision of the future that we get from big budget sci-fi films takes some kind of ecological and social apocalypse for granted. It's almost as if we have collectively stopped believing that the future is going to happen, let alone that it could be better, in many ways, than the present.

Not all art takes collapse for granted, however. There have long been creators on the margins, from Afrofuturists to feminist fantasists, who have attempted to explode the idea that the future has to be like the present, only worse and with sex robots. One such visionary was the great science-fiction writer Ursula K. Le Guin, who delivered a searing speech upon receiving the National Book Foundation Medal in 2014, four years before her death. "Hard times are coming," she said,

> when we'll be wanting the voices of writers who can see alternatives to how we live now, can see through our fear-stricken society and its obsessive technologies to other ways of being, and even imagine real grounds for hope. We'll need writers who can remember freedom—poets, visionaries—realists of a larger reality. . . . We live in capitalism, its power seems inescapable—but then, so did the divine right of kings. Any human power can be resisted and changed by human beings. Resistance and change often begin in art.

The power of art to inspire transformation is one of the original New Deal's most lasting legacies. And interestingly, back in the 1930s, that transformational project was also under relentless attack in the press, and yet it didn't slow it down for a minute.

From the start, elite critics derided FDR's plans as everything from creeping fascism to closet communism. In the 1933 equivalent of "They're coming for your hamburgers!" Republican senator Henry D. Hatfield of West Virginia wrote to a colleague, "This is despotism, this is tyranny, this is the annihilation of liberty. The ordinary American is thus reduced to the status of a robot." A former DuPont executive complained that with the government offering decent-paying jobs, "five negroes on my place in South Carolina refused work this spring . . . and a cook on my houseboat in Fort Myers quit because the government was paying him a dollar an hour as a painter."

Far-right militias formed; there was even a sloppy plot by a group of bankers to overthrow FDR.

Self-styled centrists took a more subtle tack: In newspaper editorials and op-eds, they cautioned FDR to slow down and scale back. Historian Kim Phillips-Fein, author of *Invisible Hands: The Businessmen's Crusade Against the New Deal*, told me that the parallels with today's attacks on the Green New Deal in outlets like the *New York Times* are obvious. "They didn't outright oppose it, but in many cases, they would argue that you don't want to make so many changes at once, that it was too big, too quick. That the administration should wait and study more."

And yet for all its many contradictions and exclusions, the New Deal's popularity continued to soar, winning Democrats a bigger majority in Congress in the midterms and FDR a landslide reelection in 1936.

The main reason that the elite attacks never succeeded in turning the public against the New Deal was that its programs were helping people. But another reason had to do with the incalculable power of art, which was embedded in virtually every aspect of the era's transformations. The New Dealers saw artists as workers like any other: people who, in the depths of the Depression, deserved direct government assistance to practice their trade. As Works Progress Administration director Harry Hopkins famously put it, "Hell, they've got to eat just like other people."

Through programs that included the Federal Art Project, Federal Music Project, Federal Theatre Project, and Federal Writers Project (all part of the WPA), as well as the Treasury Section of Painting and Sculpture and several others, tens of thousands of painters, musicians, photographers, playwrights, filmmakers, actors, authors, and a huge array of craftspeople found meaningful work, with unprecedented support going to African American and Indigenous artists.

The result was an explosion of creativity and a staggering body of work. The Federal Art Project alone produced nearly 475,000 works of visual art, including more than 2,000 posters, 2,500 murals, and 100,000 canvases for public spaces. Its stable of artists included Jackson Pollock and Willem de Kooning. Authors who participated in the Federal Writers' Project included Zora Neale Hurston, Ralph Ellison, and John Steinbeck. The Federal Music Project was responsible for 225,000 performances, reaching some 150 million Americans.

Much of the art produced by New Deal programs was simply about bringing joy and beauty to Depression-ravaged people—while challenging the prevalent idea that art belonged exclusively to the wealthy. As FDR put it in a 1938 letter to author Hendrik Willem van Loon, "I, too, have a dream—to show people in the out of the way places, some of whom are not only in small villages but in corners of New York City . . . some real paintings and prints and etchings and some real music."

Some New Deal art set out to mirror a shattered country back to itself and, in the process, make an unassailable case for why New Deal relief programs were so desperately needed. The result was iconic work, from Dorothea Lange's photography of Dust Bowl families enveloped in clouds of filth and forced to migrate, to Walker Evans's harrowing images of tenant farmers that filled the pages of the 1941 book *Let Us Now Praise Famous Men*, to Gordon Parks's pathbreaking photography of daily life in Harlem.

Other artists produced more optimistic, even utopian creations, using graphic art, short films, and vast murals to document the transformation under way under New Deal programs—the strong bodies building new infrastructure, planting trees, and otherwise picking up the pieces of their nation.

Just as Crabapple and I started mulling over the idea of a Green New Deal short film, inspired by the utopian art of the New Deal, *The Intercept* published a piece by Kate Aronoff that was set in the year 2043, after the Green New Deal had come to pass. It told the story of what life was like for a fictionalized "Gina," who grew up in the world that Green New Deal policies had created: "She had a relatively stable childhood. Her parents availed themselves of some of the year of paid family leave they were entitled to, and after that she was dropped off at a free child care program." After free college, "she spent six months restoring wetlands and another six volunteering at a day care much like the one she had gone to."

The piece struck a nerve, in large part because it imagined a future tense that wasn't some version of *Mad Max* warriors battling prowling bands of cannibal warlords. Crabapple and I decided that our film could do something similar, but this time from Ocasio-Cortez's vantage point. It would tell the story of how society decided to go bold rather than give up, and paint a picture of the world after the Green New Deal the congresswoman had championed became reality.

The final result is a seven-minute postcard from the future, codirected by Crabapple's longtime collaborators Kim Boekbinder and Jim Batt, and cowritten by Ocasio-Cortez and filmmaker and climate justice organizer Avi Lewis (who also happens to be my husband). It's a story about how, in the nick of time, a critical mass of humanity in the largest economy on earth came to believe that we were actually worth saving.

Crabapple's paintbrushes depict a country both familiar and entirely new. Cities are connected by bullet trains, Indigenous elders help young people restore wetlands, millions find jobs retrofitting low-cost housing—and when superstorms drown major cities, the residents respond not with vigilantism and recrimination but with cooperation and solidarity. Over those lush paintings, Ocasio-Cortez's voice is heard:

> As we battled the floods, fires and droughts, we knew how lucky we were to have started acting when we did. And we didn't just change the infrastructure. We changed how we did things. We became a society that was not only modern and wealthy, but dignified and humane, too. By committing to universal rights like health care and meaningful work for all, we stopped being so scared of the future. We stopped being scared of each other. And we found our shared purpose.

The response was unlike any we were prepared for. The film went online on April 17. Within forty-eight hours, it had been viewed well over six million times. Within seventy-two hours it was being screened in rooms of more than a thousand people, as part of a national tour to build momentum for the Green New Deal organized by the Sunrise Movement. In the halls, people cheered for every other line. Within a week, we had heard from multiple teachers (from primary through university) who told us they had already showed it in class.

"Our students are hungry for hope," read a typical report. Hundreds of people wrote to us and told us they had wept at their desks—for everything that was already lost and for everything that could still be won.

Looking back on this project, and the speed with which it traveled through the world, it strikes me that we are starting to see the true power of framing our collective response to climate change as a "Green New Deal," despite all the limitations of that historical analogy. By evoking FDR's real-world industrial and social transformation from nearly a century ago in order to imagine our world a half century from now, all of our time horizons are being stretched.

Suddenly we are no longer prisoners of the never-ending present in our social media feeds. We are part of a long and complex collective story, one in which human beings are not one set of attributes, fixed and unchanging, but rather, a work in progress, capable of deep change. By looking decades backward and forward simultaneously, we are no longer alone as we confront our weighty historical moment. We are surrounded both by ancestors whispering that we can do what our moment demands just as they did, and by future generations shouting that they deserve nothing less.

As much as the hopeful vision of the future presented by the Green New Deal, I think this lengthened time horizon is what many are responding to so powerfully. Because there is nothing more disorienting than finding yourself floating through time, unmoored from both future and past. Only when we know where we have come from, and where we want to go, will we have a sturdy place to plant our feet.

Only then will we believe, as Ocasio-Cortez says in the film, that our future has not yet been written and "we can be whatever we have the courage to see."

EPILOGUE: THE CAPSULE CASE FOR A GREEN NEW DEAL

CRITICS OF THE GREEN NEW DEAL HAVE PLENTY OF SERIOUS ARGUMENTS FOR why all this is doomed. Political paralysis in Washington is real. Even in a world where climate change–denying Republicans were swept out of power, there would still be plenty of centrist Democrats convinced that their constituents had no appetite for radical change. The plans are expensive, and getting the budgets approved would be a herculean effort.

A better course of action, we hear, would be to advance climate policies that appeal to many on the right, like a shift from coal to nuclear power, or a small tax on carbon that returns the revenues as a "dividend" to every citizen.

The main trouble with these incremental approaches is that they simply won't get the job done. In order to win support from Republicans soaked in fossil fuel money, the price on carbon would be too low to make much of an impact. Nuclear power is expensive and slow to roll out compared with renewables—and that is not to mention the risks associated with uranium mining and waste storage.

The truth is, we cannot lower emissions as steeply and as rapidly as required to swerve off our perilous trajectory without a sweeping industrial and infrastructure overhaul. The good news is that the Green New Deal isn't nearly as impractical or unrealistic as its many critics claim. I have made the case for why that is throughout this book, but what follows are nine more reasons the Green New Deal has a fighting chance—a chance that will increase every time we go out and make the case.

1. IT WILL BE A MASSIVE JOB CREATOR

Every part of the world that has invested heavily in renewables and efficiency has found these sectors to be much more powerful job creators than fossil fuels. When New York State made a commitment to get half its energy from renewables by 2030 (not fast enough), it immediately saw a spike in job creation.

The accelerated time line of the US Green New Deal will turn it into a jobs machine. Even without federal support—indeed, with active sabotage from the White House—the green economy is already creating many more jobs than oil and gas. According to the 2018 US Energy and Employment Report (USEER), jobs in wind, solar energy efficiency, and other clean energy sectors outnumbered fossil jobs by a rate of three to one. That is happening because of a combination of state and municipal incentives and the plummeting costs of renewables. A Green New Deal would take the industry supernova while ensuring that the jobs have salaries and benefits comparable to those offered in the oil and gas sector.

There is no shortage of research to support this. For instance, a 2019 study on the job impacts of a Green New Deal–style program in the state of Colorado found that many more jobs would be created than lost. The study, published by the Department of Economics and Political Economy Research Institute at University of Massachusetts–Amherst, looked at what it would take for the state to achieve a 50 percent reduction in emissions by 2030. It found that roughly 585 nonmanagement jobs would be lost but that, with an investment of $14.5 billion a year in clean energy, "Colorado will generate about 100,000 jobs per year in the state."

There are many more studies with similarly striking findings. A plan put forward by the U.S. BlueGreen Alliance, a body that brings together unions and environmentalists, estimated that a $40 billion annual investment in public transit and high-speed rail for six years would produce more than 3.7 million jobs during that period. And according to a report for the European Transport Workers Federation, comprehensive policies to reduce emissions in the transport sector by 80 percent would create seven million new jobs across the continent, while another five million clean energy jobs in Europe could slash electricity emissions by 90 percent.

2. PAYING FOR IT WILL CREATE A FAIRER ECONOMY

As the 2018 IPCC report on keeping warming below 1.5°C made clear, if we don't take transformative action to lower emissions, the costs will be astronomical. The panel's estimate is that the economic damages of allowing temperatures to increase by 2°C (as opposed to 1.5°C) would hit $69 trillion globally.

Of course, rolling out a Green New Deal would have large costs as well, and the plan's advocates have pointed to a variety of ways this can be financed. Alexandria Ocasio-Cortez has said that the US version should be financed the way any previous emergency spending has been: by the US Congress simply authorizing the funds, backstopped by the Treasury of the world's currency of last resort. According to New Consensus, the think tank closely associated with her policy proposals, because "the Green New Deal will produce new goods and services to keep pace with and absorb new expenditures, there is no more reason to let fear about financing halt progress here than there was to let it halt wars or tax cuts."

The European Spring proposal for a Green New Deal, meanwhile, calls for a global minimum corporate tax rate to capture the tax revenue that the Apples and Googles of the world currently dodge with transnational schemes. It also calls for a reversal of monetary orthodoxy, with public investment floating green bonds, supported by central banks. "To address the true existential threat that we face today, we must reverse the economic policies that brought us to this brink. Austerity means extinction." Some analysts, like Christian Parenti, have emphasized that federal governments can drive the transition with their purchasing policies.

In short, there are all kinds of ways to raise financing, including ways that attack untenable levels of wealth concentration and shift the burden to those most responsible for climate pollution. And it's not hard to figure out who that is. We know, thanks to research from the Climate Accountability Institute, that a whopping 71 percent of total greenhouse gas emissions since 1988 can be traced to just one hundred corporate and state fossil fuel giants, dubbed the "Carbon Majors."

In light of this, there are a variety of "polluter pays" measures that can be taken to ensure that those most responsible for this crisis do the most to underwrite the transition—through legal damages, through higher royalties, and by having their subsidies slashed. Direct fossil fuel subsidies are worth about $775 billion a year globally, and more than $20 billion in the United States alone.

The very first thing that should happen is these subsidies should be shifted to investments in renewables and efficiency.

It isn't only the fossil fuel companies who have put their own super profits ahead of the safety of our species for decades; so have the financial institutions that underwrote their investments, in full knowledge of the risks. Which is why, in addition to eliminating fossil fuel subsidies, governments can insist on getting a much fairer share of the financial sector's massive earnings by imposing a transaction tax, which could raise $650 billion globally, according to the European Parliament.

And then there is the military. If the military budgets of the top ten military spenders globally were cut by 25 percent, that would free up $325 billion annually, according to numbers reported by the Stockholm International Peace Research Institute—funds that could be spent on energy transition and preparing communities for the extreme weather ahead.

Meanwhile, a mere 1 percent billionaire's tax could raise $45 billion a year globally, according to the United Nations—not to mention the money that would be raised through an international effort to close down tax havens. In 2015 the estimated private financial wealth of individuals stashed unreported in tax havens around the globe was somewhere between $24 trillion and $36 trillion, according to James S. Henry, a senior adviser with the UK-based Tax Justice Network. Shutting down some of those havens would go a very long way toward covering the price tag of this desperately needed industrial transition.

3. IT TAPS THE POWER OF EMERGENCY

A Green New Deal approach does not treat the climate crisis as one issue on a checklist of worthy priorities. Rather, it heeds Greta

Thunberg's call to "act like your house is on fire. Because it is." The truth is that the scientific deadline for deep transformation is so short that if radical change doesn't roll out every year for the next thirty years, we will have lost the tiny window we have to avert truly catastrophic warming. Treating an emergency like an emergency means all our energies can go into action, rather than into screaming about the need for action, which what is happening now.

That, in turn, would liberate us all from the debilitating cognitive dissonance that living in a culture that is denying the reality of so profound a crisis requires. The Green New Deal puts us all on emergency footing: as scary as that would be for some, the catharsis and relief for many others, particularly young people, would be a potent source of energy.

4. IT'S PROCRASTINATION-PROOF

Some have criticized the Green New Deal resolution for stating that the United States must get off fossil fuels in just a decade. Scientists have said the world needs to get to net-zero emissions by 2050, so why the rush? The first answer is "justice": wealthy countries that became that way by polluting without limit need to decarbonize fastest, so that poorer countries where majorities still lack the basics of clean water and electricity can have a more gradual transition.

But the second answer is strategic: a ten-year deadline means there can be no more procrastination. Up until the Green New Deal, every political response to the climate crisis set the most ambitious targets decades in the future, long after the politicians making these pledges would leave office. The tasks these politicians gave to themselves, though, were in comparison relatively easy, like introducing cap-and-trade schemes or retiring old coal plants and replacing them with natural gas. The tough work of confronting the fossil fuel industry's entire business model was perennially offloaded onto successors.

Embracing a ten-year-transition time line does not mean absolutely everything has to get done in a decade. The resolution sets an ambitious deadline but repeatedly adds "to the extent technologically feasible." Fundamentally, this means that we are no longer kicking the can down the road. The current crop of politicians introducing a Green New Deal would finally be saying, "We are the ones who are going to get the job done. Not someone else."

Given the damage that the temptation to procrastinate has already done to our planet, that is a very big deal.

5. IT'S RECESSION-PROOF

Over the past three decades, one of the greatest obstacles to making sustained progress on climate action has been the volatility of the market. During good economic times, there is usually some willingness to entertain environmental policies that mean paying a bit more for gas, electricity, and "green" products. But again and again, this willingness has understandably evaporated as soon as the economy hit a painful downturn.

And that may be the greatest benefit of modeling our climate approach after FDR's New Deal, the most famous economic stimulus of all time, one born in the teeth of the worst economic crisis in modern history. When the global economy enters another downturn or crisis, which it surely will, support for a Green New Deal will not plummet as has been the case with every other major green initiative during past recessions. Instead, support can be expected to increase, since a large-scale stimulus with the power to create millions of jobs will become the greatest hope of addressing people's economic pain.

6. IT'S A BACKLASH BUSTER

Too often, when politicians introduce climate policies divorced from a broader agenda of economic justice, the policies they introduce are actively unjust—and the public responds accordingly. Look, for example, at France under Emmanuel Macron, derided by his opponents as the "president for the rich." Macron has pursued a classic "free-market" agenda for France, cutting taxes for the wealthy and corporations, rolling back hard-won labor protections, making higher education less accessible—all after years of austerity under previous administrations.

It was in this context that, in 2018, he introduced a fuel tax designed to make driving move expensive, thereby reducing consumption and raising some funds for climate programs.

Except it didn't work like that. Huge numbers of working people in France, already under intense economic stress from Macron's other policies, saw this market-based approach to the climate crisis as a direct attack on them: Why should they pay more to drive themselves to work when the super-rich were free to fuel up their private jets to visit their tax havens? Tens of thousands took to the streets in anger, many of them wearing yellow safety vests (*gilets jaunes*), with several protests turning into full-scale riots.

"The government cares about the end of the world," many *gilets jaunes* chanted. "We care about the end of the month." Desperately trying to regain control over the country, Macron rolled back his fuel tax and introduced a minimum wage hike, among other concessions—at the same time that he brutally repressed the movement.

One of the great strengths of a Green New Deal approach is that it will not generate this kind of backlash. Nothing about its framework forces people to choose between caring about the end of the world and the end of the month. The whole point is to design policies that allow us all to care about both, policies that simultaneously lower emissions and lower the economic strain on working people—by making sure that everyone can get a good job in the new economy; that they have access to basic social protections like health care, education, and daycare; and that green jobs are good, unionized, family-supporting jobs with benefits and vacation time. There will certainly have to be a price on carbon, but it has a much better chance of survival if the people who pay the increased costs aren't hanging on by their fingernails.

7. IT CAN RAISE AN ARMY OF SUPPORTERS

Since its launch, the most frequent criticism of the Green New Deal is that, by focusing so much on economic and social justice, it is making climate action a harder sell than a plan more narrowly trained on carbon emissions. "My heart is with the greens," Thomas Friedman wrote in the *New York Times*. "But my head says you can't transform our energy system and our social/economic one at scale all at once. We have to prioritize energy/climate. Because for the environment, later will be too late. Later is officially over."

This assumes that the social/economic components of the Green New Deal are weighing it down. In fact, they are precisely what is lifting it up.

Unlike approaches that pass on the costs of transition to working people, the Green New Deal is squarely focused on marrying pollution reduction to the top priorities of the most vulnerable workers and the most excluded communities. This is the game changer of having representatives in Congress rooted in working-class struggles for living wage jobs and for nontoxic air and water—women like Rashida Tlaib, who helped fight a successful battle against Koch Industries' noxious petroleum coke mountain in Detroit.

If you are part of the economy's winning class and funded by even bigger winners, as so many politicians are, then your attempts to craft climate legislation tend to be guided by the idea that change should be as minimal and as unchallenging to the status quo as possible. After all, the status quo is working just fine for you and your donors. That was the approach that failed to get cap-and-trade through the Senate during the Obama years, and it's the approach that blew up in Macron's face in France.

Leaders who are rooted in communities that are being egregiously failed by the current system, on the other hand, are liberated to take a very different approach. Their climate policies can embrace deep and systemic change because deep change is precisely what their bases need to thrive.

For decades, the biggest barrier to winning climate legislation has been a vast power mismatch. Opposition to action from fossil fuel companies was ferocious, creative, and tenacious. But when it came to the kinds of weak (and very often unjust) market-based climate policies that made it onto the political agenda, support was tepid at best.

The Green New Deal, however, is already showing that it has the power to mobilize a truly intersectional mass movement behind it—not despite its sweeping ambition, but precisely because of it. As climate justice organizations have been arguing for many years now, when the communities with the most to gain from change lead the movement, they fight to win.

8. IT WILL BUILD NEW ALLIANCES—AND UNDERCUT THE RIGHT

One of the knocks on the Green New Deal is that, by linking climate action with so many other progressive policy goals, conservatives will be more convinced that global warming is a plot to smuggle in socialism, and political polarization will deepen.

There is no question that Republicans in Washington will continue to paint the Green New Deal as a recipe for turning the United States into Venezuela—about that we can all be certain. But this worry misses one of the greatest benefits of approaching the climate emergency as a vast infrastructure and land regeneration project: nothing heals ideological divides faster than a concrete project bringing jobs and resources to hurting communities.

One person who understood this well was FDR. When he rolled out the network of camps that made up the Civilian Conservation Corp, for example, he purposely clustered many of them in rural areas that had voted against him for president. Four years later, when those communities had experienced the benefits of the New Deal up close, they were far less vulnerable to Republican fearmongering about a socialist takeover of government, and many voted Democrat.

We can expect a massive rollout of job-creating green infrastructure and land rehabilitation projects to have a similar effect today. Some people will still be convinced that climate change is a hoax—but if it's a hoax that creates good jobs and detoxifies the environment, especially in regions where the only other economic development on offer is a supermax prison, who really cares?

9. WE WERE BORN FOR THIS MOMENT

By far the biggest obstacle we are up against is hopelessness, a feeling that it's all too late, we've left it too long, and we'll never get the job done on such a short time line.

And all of that would be true if the process of transformation were starting from scratch. But the truth is that there are tens of thousands of people, and a great many organizations, who have been preparing for a Green New Deal–style breakthrough for decades (centuries in the case of Indigenous communities that have been protecting their ways of life). These forces have been quietly building local models and road testing policies for how to put justice at the center of our climate response—in how we protect forests, generate renewable energy, design public transit, and much more.

"Who is society?" demanded then–British prime minister Margaret Thatcher in 1987, justifying her relentless attacks on social services. "There is no such thing! There are individual men and women and there are families."

That bleak view of humanity—that we are nothing more than a collection of atomized individuals and nuclear families, unable to do anything of value together except wage war—has had a stranglehold over the public imagination for a very long time. No wonder so many of us believed we could never rise to the climate challenge.

But more than thirty years later, as surely as the glaciers are melting and the ice sheets are breaking apart, that "free-market" ideology is dissolving, too. In its place, a new vision of what humanity can be is emerging. It is coming from the streets, from the schools, from workplaces, and even from inside houses of government. It's a vision that says that all of us, combined, make up the fabric of society.

And when the future of life is at stake, there is nothing we cannot achieve.

ACKNOWLEDGMENTS

JONATHAN KARP AT SIMON & SCHUSTER SAW THIS BOOK THROUGH FROM conception to publication. I am grateful for his faith in me, for his many helpful editorial insights, and for his sense of urgency about the state of our world. Lake Bunkley helped us every step of the way and Jenna Dolan provided a careful copyedit. I am delighted to be working once again with my longtime editor, collaborator, and dear friend Louise Dennys at Penguin Random House Canada, who provided many insightful notes. We are all pleased to be working with the dedicated team at Penguin Random House UK.

My friend Anthony Arnove acted as my agent, finding happy homes for this book in translation around the world, as well as providing valuable editorial comments. Jackie Joiner kept all the moving pieces humming along, from website launch to tour planning. I would be lost without her colleagueship. We are both grateful to Julia Prosser, Shona Cook, Annabel Huxley, and too many others to name here.

Rajiv Sicora, my brilliant collaborator since *This Changes*

Everything, assisted with research for most of the essays in this book. Sharon J. Riley was the researcher on "Season of Smoke" and "The Leap Years: Ending the Story of Endlessness." Jennifer Natoli and Nicole Weber helped tremendously with new research and updates. I am grateful to Eyal Weizman for allowing me to use his map of the aridity line, as it appears in *Forensic Architecture*.

Johann Hari was a wonderfully astute first reader and kind friend. I am also grateful to the original editors on these pieces, especially my longtime colleagues and friends: Betsy Reed, Roger Hodge, Richard Kim, and Katharine Viner. I could not walk this road without the wisdom and support of other friends and family members including Kyo Maclear, Bill McKibben, Eve Ensler, Nancy Friedland, Andréa Schmidt, Astra Taylor, Keeanga-Yamahtta Taylor, Harsha Walia, Molly Crabapple, Janice Fine, Seumas Milne, Jeremy Scahill, Cecilie Surasky, Melina Laboucan-Massimo, Bonnie Klein, Michael Klein, Seth Klein, Misha Klein, Christine Boyle, Michele Landsberg, and the indomitable Stephen Lewis. Courtney Butler and Fatima Lima protected space for me to work.

I have been supported throughout this writing by my new community of colleagues at Rutgers University, including Jonathan Potter, Dafna Lemish, Juan Gonzáles, Mary Chayko, Lisa Hetfield, and especially Kylie Davidson. I am indebted to Gloria Steinem whose lifelong work created the position I now hold. I am grateful to everyone at *The Intercept* for engaging in fearless journalism and providing such a welcome home for my work; same goes for Type Media Institute (formerly The Nation Institute) where I am a fellow. The kick-ass team at The Leap works all day, every day to turn the vision in these pages into a lived reality for all of us. They inspire me beyond measure, and our management team of Leah Henderson, Katie McKenna, and Bianca Mugyenyi lead us with unflagging ambition and confidence.

This book is dedicated to the memory of my friend and teacher Arthur Manuel, whose absence has left an unfillable hole in my life and in the global movements for climate justice and Indigenous sovereignty. I am grateful to the members of the entire Manuel family who keep his legacy alive and show us all what true leadership looks like.

My husband, Avi Lewis, provided stellar editorial advice, as he has for every book I have written. The fact that he was busily making films about the Green New Deal and helping to get new Green New Deal coalitions off the ground in more than one country was extremely distracting. Our son, Toma, reminds us both daily that failure is not an option.

PUBLICATION CREDITS

Versions of several of these essays appeared in modified forms under the following titles:

"Gulf Oil Spill: A Hole in the World," *The Guardian*, June 18, 2010.

"Capitalism vs. the Climate," *The Nation*, November 9, 2011.

"Geoengineering: Testing the Waters," *New York Times*, October 27, 2012.

"How Science Is Telling Us All to Revolt," *New Statesman America*, October 29, 2013.

"Climate Change Is the Fight of Our Lives—Yet We Can Hardly Bear to Look at It," *The Guardian*, April 23, 2014.

"Climate Change Is a Crisis We Can Only Solve Together," College of the Atlantic 2015 Commencement Address, Bar Harbor, ME, June 6, 2015.

"A Radical Vatican," *The New Yorker*, July 10, 2015.

"Let Them Drown: The Violence of Othering in a Warming World," 2016 Edward W. Said London Lecture, April 5, 2016. Published in the *London Review of Books*, June 2, 2016.

"We Are Hitting the Wall of Maximum Grabbing," 2016 Sydney Peace Prize Acceptance Speech, November 11, 2016. Published in *The Nation*, December 14, 2016.

"Season of Smoke: In a Summer of Wildfires and Hurricanes, My

Son Asks 'Why Is Everything Going Wrong?'" *The Intercept*, September 9, 2017, Research assistance: Sharon J. Riley.

"Speech to the 2017 Labour Party Conference," 2017 Labour Party Annual Conference, Brighton, UK, September 26, 2017.

"Capitalism Killed Our Climate Momentum, Not 'Human Nature,'" *The Intercept*, August 3, 2018.

"There's Nothing Natural About Puerto Rico's Disaster," *The Intercept*, September 21, 2018. Piece based on remarks given September 20 at "One Year Since Maria," a rally in Union Square Park in New York City, organized by UPROSE and OurPowerPRnyc.

"The Battle Lines Have Been Drawn on the Green New Deal," *The Intercept*, February 13, 2019.

INDEX